Collins

KS3
Maths
Foundation Level

Revision Guide

Samya Abdullah, Rebecca Evans,
Trevor Senior and Gillian Spragg

About this Revision & Practice book

When it comes to getting the best results, practice really does make perfect!

Experts have proved that repeatedly testing yourself on a topic is far more effective than re-reading information over and over again. And, to be as effective as possible, you should space out the practice test sessions over time.

This Complete Revision & Practice book is specially designed to support this approach to revision and includes seven different opportunities to test yourself on each topic, spaced out over time.

This book is suitable for students working upto the expected level in Key Stage 3 Maths.

Symbols are used to highlight questions that test key skills:

(MR) Mathematical Reasoning (PS) Problem Solving

(FS) Financial Skills

Try to answer as many questions as possible without using a calculator. Questions where calculators **must not** be used are marked with this symbol:

Revise

These pages provide a recap of everything you need to know for each topic. All key words are defined in the glossary.

You should read through all the information before taking the Quick Test at the end. This will test whether you can recall the key facts.

> **Quick Test**
>
> 1. Simplify $4x + 7y + 3x - 2y + 6$
> 2. Simplify $c \times c \times d \times d$
> 3. Find the value of $4x + 2y$ when $x = 2$ and $y = 3$

Practise

These topic-based questions appear shortly after the revision pages for each topic and will test whether you have understood the topic. There are two levels of demand for each topic.

Review

These topic-based questions appear later in the book, allowing you to revisit the topic and test how well you have remembered the information. There are two levels of demand for each topic.

Mix it Up

These pages feature a mix of questions for all the different topics, just like you would get in a test. They will make sure you can recall the relevant information to answer a question without being told which topic it relates to.

Test Yourself on the Go

Visit our website at **collins.co.uk/collinsks3revision** and print off a set of flashcards. These pocket-sized cards feature questions and answers so that you can test yourself on all the key facts anytime and anywhere. You will also find lots more information about the advantages of spaced practice and how to plan for it.

Workbook

This section features even more topic-based questions (again with two levels of demand) and mixed test-style questions, providing two further practice opportunities for each topic to guarantee the best results.

ebook

To access the ebook, visit

collins.co.uk/ebooks

and follow the step-by-step instructions.

QR Codes

Found throughout the book, the QR codes can be scanned on your smartphone for extra practice and explanations.

A QR code in the Revise section links to a Quick Recall Quiz on that topic. A QR code in the Workbook section links to a video working through the solution to one of the questions on that topic.

Contents

		Revise	Practise	Review

N Number   **A** Algebra   **G** Geometry and Measures

S Statistics   **P** Probability   **R** Ratio, Proportion and Rates of change

Contents

N Number   **A** Algebra   **G** Geometry and Measures

S Statistics   **P** Probability   **R** Ratio, Proportion and Rates of change

Contents

Review Questions

Key Stage 2: Key Concepts

1   Which of these two numbers is closer to 2000?

1996 or 2007

Explain how you know. [2]

2   Calculate 476 – 231 📱 [2]

3   Copy and complete the table below by rounding each number to the nearest 1000

	To the nearest 1000
4587	
45 698	
457 658	
45 669	

[2]

4   Write these values in order, starting with the smallest.

0.56   55% $\frac{27}{50}$ 0.6   0.63 [3]

5   Ahmed is twice as old as Rebecca.

Rebecca is three years younger than John.

John is 25 years old.

How old is Ahmed? [2]

6   Calculate 467 × 34 📱 [3]

7   Calculate 156 ÷ 3 📱 [2]

8   On the scale below, draw arrows to show 1.6 and 3.8

[2]

1 A bottle holds 1 litre of fizzy drink. Mariam pours four glasses for her friends.

Each glass contains 200 ml.

How much fizzy drink is left in the bottle? [3]

2 Below are five digit cards.

| 7 | 5 | 1 | 6 | 3 |

Choose two cards to make the following two-digit numbers.

a) A square number [1]

b) A prime number [1]

c) A multiple of 6 [1]

d) A factor of 60 [1]

3 An equilateral triangle has a perimeter of 27 cm.

What is the length of one of its sides? [2]

4 Two-thirds of a number is 22

What is the number? [2]

5 Here is an isosceles triangle drawn inside a rectangle.

Find the value of the angle x. [3]

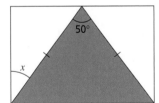

6 S and T are two whole numbers.

$S + T = 500$

S is 100 greater than T.

Find the value of S and T. [2]

Total Marks .... / 16

Number 1

You must be able to:

- Order positive and negative numbers
- Carry out additions and subtractions involving negative numbers
- Use the symbols $=, \neq, <, >, \leq, \geq$
- Multiply and divide integers
- Carry out operations following BIDMAS.

Positive and Negative Numbers

- A **number line** can be used to order **integers**.

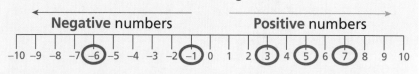

Place 7, 5, –6, –1 and 3 in ascending order.

Going from left to right, you can see that the ascending order is –6, –1, 3, 5, 7

> **Key Point**
>
> There is an infinite number of positive and negative numbers.

- You can use a number line to add and subtract numbers.

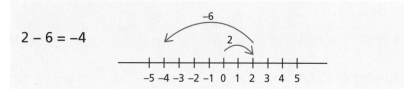

$2 - 6 = -4$

- **Place value** can be used to compare the size of large numbers.

Which is greater, 3408 or 3540?

Number	Thousands	Hundreds	Tens	Units
3408	3	4	0	8
3540	3	5	4	0

Both numbers have 3 thousands, but 3540 has 5 hundreds and 3408 only has 4 hundreds. Therefore 3540 is greater than 3408.

> **Key Point**
>
> Always compare digits from left to right.

- **Symbols** are used to state the relationship between two numbers.

Symbol	Meaning	Symbol	Meaning
$=$	Equal to	$\neq$	Not equal to
$<$	Less than	$\leq$	Less than or equal to
$>$	Greater than	$\geq$	Greater than or equal to

3540 is greater than 3408 can be written as $3540 > 3408$
–10 is less than –5 can be written as $-10 < -5$
$-6 < -2$ $3 > -1$ $2 + 3 \neq 23$

Multiplication and Division

- To multiply large numbers, you can use the **grid** (or **box**) method or a **column** method.

Calculate 354×273

Grid or Box method

×	300	50	4	
200	60 000	10 000	800	70 800
70	21 000	3500	280	24 780
3	900	150	12	1 062
				96 642

$354 = 300 + 50 + 4$
$273 = 200 + 70 + 3$

Adding the numbers in the grid gives $354 \times 273 = 96\ 642$

Column method

$$
\begin{array}{r}
3\ \ 5\ \ 4 \\
\times \quad 2\ \ 7\ \ 3 \\
\hline
1\ \ 0\ ^1 6\ ^1 2 \quad (3 \times 354) \\
2\ \ 4\ ^3 7\ ^2 8\ \ 0 \quad (70 \times 354) \\
7\ ^1 0\ \ 8\ \ 0\ \ 0 \quad (200 \times 354) \\
\hline
9\ \ 6\ ^1 6\ ^1 4\ \ 2
\end{array}
$$

So, $354 \times 273 = 96\ 642$

- Division can also be broken down into steps.

Calculate $762 \div 3$

Short division

$3\overline{)7\ 6\ 2}$	$3\overline{)7\ ^1 6\ 2}$	$3\overline{)7\ ^1 6\ ^1 2}$	$3\overline{)7\ ^1 6\ ^1 2}$
Set up the division.	$7 \div 3 = 2$ remainder 1	$16 \div 3 = 5$ remainder 1	$12 \div 3 = 4$ remainder 0

So, $762 \div 3 = 254$

Long division

$$
\begin{array}{r}
2\ 5\ 4 \\
3\overline{)7\ 6\ 2} \\
6 \\
\hline
1\ 6 \\
1\ 5 \\
\hline
1\ 2 \\
1\ 2 \\
\hline
0
\end{array}
$$

BIDMAS

- **BIDMAS** gives the order in which operations should be carried out:

  Brackets
  Indices (powers)
  Division *and*
  Multiplication
  Addition *and*
  Subtraction

$2 \times 6 + 5 \times 4 = 32$
$2 \times 6 + 5 \times 4 + 0 = 32$

Adding 0 does not change the answer. 0 is called an **additive identity** as it leaves the answer unchanged.

$(3 + 4) \times (9 - 1) = 7 \times 8 = 56$
$(3 + 4) \times (9 - 1) \times 1 = 7 \times 8 \times 1 = 56$

Multiplying by 1 does not change the answer. 1 is called a **multiplicative identity** as it leaves the answer unchanged.

Key Point

Division and multiplication should be carried out in the order they appear in the calculation from the left. So carry out the multiplication first if it appears before the division.

Addition and subtraction should be carried out in the order they appear in the calculation from the left. So carry out the subtraction first if it appears before the addition.

Key Words

integer
positive
negative
place value
BIDMAS

Quick Test

1. Put 3750, 3753, 3601, 3654 and 3813 in ascending order.
2. Work out 435×521
3. Work out $652 \div 4$
4. Work out $4 \times 3 + 7 \times 4$
5. Write the following using a mathematical symbol: -5 is less than 3

Number 2

You must be able to:

- Understand square numbers, square roots, cube numbers and cube roots
- Write a number as a product of prime factors
- Find the lowest common multiple and highest common factor.

Quick Recall Quiz

Squares, Square Roots, Cubes and Cube Roots

- **Square numbers** are calculated by multiplying a number by itself.

$$5^2 = 5 \times 5 = 25$$

- A **square root** ($\sqrt{}$) is the inverse or opposite of a square.

$\sqrt{36} = 6$ (because $6 \times 6 = 36$) $\sqrt{25} = 5$ (because $5 \times 5 = 25$)

$\sqrt{30}$ is between 5 and 6 $\sqrt{30} = 5.47...$

- **Cube numbers** are calculated by multiplying a number by itself and by itself again.

$$4^3 = 4 \times 4 \times 4 = 64 \qquad 4^2 + 3^3 = 16 + 27 = 43$$

- A **cube root** ($\sqrt[3]{}$) is the inverse or opposite of a cube.

$$\sqrt[3]{8} = 2 \text{ because } 2 \times 2 \times 2 = 8$$

- There are other powers as well as squares and cubes.

$$5^4 = 5 \times 5 \times 5 \times 5 = 625 \qquad 5^5 = 5 \times 5 \times 5 \times 5 \times 5 = 3125$$

> **Key Point**
>
> The first ten square numbers are:
>
> 1, 4, 9, 16, 25, 36, 49, 64, 81, 100

> **Key Point**
>
> The first five cube numbers are:
>
> 1, 8, 27, 64, 125

Reciprocals

- A **reciprocal** is the inverse of any number except zero, e.g. the reciprocal of 2 is $\frac{1}{2}$

The reciprocal of $\frac{1}{2}$ is $1 \div \frac{1}{2} = 2$

Prime Factors

- A **factor** of an integer is an integer that divides exactly into it without leaving a remainder. Factors are integers you can multiply together to make another number.

The factors of 12 are 1, 2, 3, 4, 6 and 12 because $1 \times 12 = 12$, $2 \times 6 = 12$ and $3 \times 4 = 12$

This is called the unique factorisation property.

- Every integer greater than 1 is either prime or can be written as the **product** of a unique list of prime numbers.
- A **prime number** has exactly two factors, itself and 1

> **Key Point**
>
> Product means multiply.

- Breaking a number down into a product of prime factors is called **prime factor decomposition**.

45 can be expressed as a product of prime factors. This can be done using a **prime factor tree**:

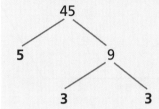

$45 = 5 \times 9$
and
$9 = 3 \times 3$
So $45 = 3 \times 3 \times 5$

Key Point

A prime factor tree breaks a number down into its prime factors.

Always start by finding a prime number which is a factor, in this case 5

Remember to write the final answer as a product.

LCM and HCF

- The **lowest common multiple** (LCM) is the lowest multiple two or more numbers have in common.
- The **highest common factor** (HCF) is the highest factor two or more numbers have in common.

Find the lowest common multiple and highest common factor of 12 and 42

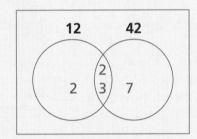

The LCM is the product of all the numbers in both circles.

LCM $= 2 \times 2 \times 3 \times 7 = 84$

The HCF is the product of the numbers in the overlap.

HCF $= 2 \times 3 = 6$

Write both numbers as a product of prime factors.
$12 = 2 \times 2 \times 3$ $42 = 2 \times 3 \times 7$
Complete a Venn diagram.

Common factors are placed in the overlap.

- LCM and HCF are used to solve many everyday problems.

Tom swims in competitions. He visits his doctor every 12 days. He visits his nutritionist every 15 days. He saw both of them on 1st October. On what date will he next see both of them on the same day?

Find the LCM of 12 and 15: this is 60

60 days after 1st October is 30th November.

Key Words

square number
square root
cube number
cube root
reciprocal
factor
product
prime
prime factor
    decomposition
lowest common multiple
highest common factor

Quick Test

1. Write down the values of: **a)** 7^2 **b)** 4^3
2. Write down the values of: **a)** $\sqrt{49}$ **b)** $\sqrt[3]{27}$
3. Write 40 as a product of prime factors.
4. Find the lowest common multiple of 14 and 36
5. Find the highest common factor of 24 and 32

Sequences 1

You must be able to:

- Use a function machine to generate terms of a sequence
- Recognise arithmetic and geometric sequences
- Generate sequences from a term-to-term rule.

Function Machines

- A **function machine** (number machine) takes an input, applies one or more operations and produces an output.
- Function machines can be shown in different ways:

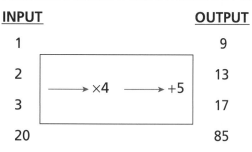

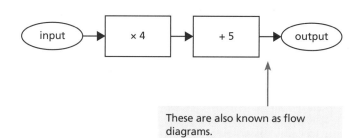

These are also known as flow diagrams.

Sequences

- A **sequence** is a set of shapes, numbers or letters which follow a pattern or rule.

The outputs from a function machine can form a sequence.

- Each part of a sequence is called a **term**.
- Any sequence that can carry on forever is called **infinite**.

In the sequence below, the next pattern is formed by adding extra tiles around the previous pattern.

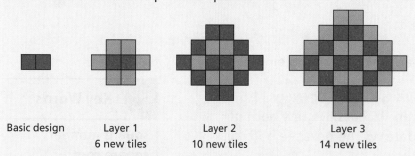

| Basic design | Layer 1
6 new tiles | Layer 2
10 new tiles | Layer 3
14 new tiles |

The number of new tiles needed increases by 4

This pattern can be used to predict how many tiles will be needed to make larger designs.

- An **arithmetic sequence** is a set of numbers with a **common difference** between consecutive terms.

  4, 7, 10, 13, 16,... is an arithmetic sequence with a common difference of 3

  12, 8, 4, 0, –4,... is also an arithmetic sequence. It has a common difference of –4

Key Point

Arithmetic sequences are used to solve many real-life problems.

- In a **geometric sequence**, each term is multiplied by the same value to find the next term.

  > 2, 4, 8, 16, 32, 64,... is a geometric sequence where each term is multiplied by 2 to find the next term.

- There are many other sequences of numbers that follow a pattern:
  - 1, 4, 9, 16, 25, 36,... are square numbers.
  - 1, 8, 27, 64, 125, 216,... are cube numbers.
  - 1, 3, 6, 10, 15, 21,... are known as the **triangular numbers**.
  - 1, 1, 2, 3, 5, 8, 13,... is known as the **Fibonacci sequence**.

  Each number is the sum of the previous two numbers.

Finding Missing Terms

- The **term-to-term rule** links each term in the sequence to the previous term.

  > 5, 8, 11, 14, 17,...
  >
  > In this set of numbers the next term is found by adding 3 to the previous term. Therefore the term-to-term rule is +3.
  >
  > This rule can be used to find the next numbers in the sequence.
  >
  > 17 + 3 = 20   20 + 3 = 23   23 + 3 = 26
  >
  > Therefore the next three terms in the sequence are 20, 23, 26,...

- The term-to-term rule can also be used to find missing terms.

  > 13, 17, 21, _____, 29,...
  >
  > The term-to-term rule is +4 and so the missing term is 25

Quick Test

1. Find the missing input and output in the function machine.

   INPUT   OUTPUT

   6   ☐
      ⟶ ÷2 ⟶ +1
   ☐   25

2. Write down the next five terms in this sequence:
   5, 9, 13, 17,...
3. Write down the term-to-term rule in this sequence.
   24, 12, 6, 3, 1.5,...
4. Which of the following sequences is arithmetic?
   A) 4, 6, 9, 13, 18,...   B) 3, 6, 9, 12, 15,...
5. Find the missing term in the following sequence of numbers.
   21, 27, 33, _____, 45, 51,...

Key Words

function machine
sequence
term
infinite
arithmetic sequence
common difference
geometric sequence
triangular numbers
term-to-term

Sequences 2

Algebra

You must be able to:

- Generate terms of a sequence from a position-to-term rule
- Find the nth term of an arithmetic sequence
- Recognise quadratic sequences.

The nth Term

- The **nth term** is also called the **position-to-term rule**.
- It is an algebraic expression that represents the operations carried out by a function machine.

INPUT		OUTPUT
1		9
2	$\longrightarrow \times 4 \longrightarrow +5$	13
3		17
n		$4n + 5$

- The **nth term** can be used to generate terms of a sequence.

> The nth term of a sequence is given by $3n + 5$
>
> To find the first term, you **substitute** $n = 1$
>
> $3 \times 1 + 5 = 8$ 8 is the first term in the sequence.
>
> To find other terms, you can substitute different values of n.
>
> | When $n = 2$ | When $n = 3$ | When $n = 4$ |
> |--------------|--------------|--------------|
> | $3 \times 2 + 5 = 11$ | $3 \times 3 + 5 = 14$ | $3 \times 4 + 5 = 17$ |
> | Second term = 11 | Third term = 14 | Fourth term = 17 |
>
> The nth term $3n + 5$ produces the sequence of numbers:
>
> 8, 11, 14, 17, 20,...
>
> The rule can be used to find any term in the sequence.
> For example, to find the 50th term in the sequence
> substitute $n = 50$
>
> $3 \times 50 + 5 = 155$

Key Point

For the first term in the sequence, n always equals 1

Finding the *n*th Term

- To find the *n*th term, look for a pattern in the sequence of numbers.

The first five terms of a sequence are 7, 11, 15, 19, 23

The term-to-term rule is +4 so the *n*th term starts with 4*n*.

The difference between 4*n* and the output in each case is 3, so the final rule is 4*n* + 3

Input	× 4	Output
1	4	7
2	8	11
3	12	15
4	16	19
5	20	23
n	4*n*	4*n* + 3

Quadratic Sequences

- **Quadratic** sequences are based on square numbers.

The first five terms of the sequence $2n^2 + 1$ are as follows:

When $n = 1$ $2 \times (1)^2 + 1 = 3$
When $n = 2$ $2 \times (2)^2 + 1 = 9$
When $n = 3$ $2 \times (3)^2 + 1 = 19$
When $n = 4$ $2 \times (4)^2 + 1 = 33$
When $n = 5$ $2 \times (5)^2 + 1 = 51$

The sequence starts 3, 9, 19, 33, 51,…

The *n*th term of a sequence is given by $\frac{n(n+1)}{2}$

Work out the first four terms of the sequence.

Substituting $n = 1, 2, 3, 4$ gives 1, 3, 6, 10

> **Key Point**
>
> Use BIDMAS when calculating terms in a sequence.

This is the sequence of triangular numbers.

> **Quick Test**
>
> 1. Write down the first five terms in the sequence 5*n* + 3
> 2. Write down the first five terms in the sequence $n^2 + 4$
> 3. a) Find the *n*th term for the following sequence of numbers:
>   6, 9, 12, 15, 18,…
>   b) Find the 50th term in this sequence.
> 4. What is the *n*th term also known as?

> **Key Words**
>
> *n*th term
> position-to-term
> substitute
> quadratic

Practice Questions

Number

1 The table shows the average daily minimum temperature and the average daily maximum temperature in a town for every month of the year (given to the nearest °C).

Month	Jan	Feb	Mar	Apr	May	Jun	Jul	Aug	Sep	Oct	Nov	Dec
Average Daily Minimum Temp.	–8	–4	–1	3	8	11	14	13	9	3	–2	–6
Average Daily Maximum Temp.	–1	4	11	17	21	25	28	27	22	13	5	0

a) In which month is the average daily minimum temperature lowest? [1]

b) In which month is the average daily maximum temperature highest? [1]

c) In March, what is the difference between the average daily minimum temperature and the average daily maximum temperature? [1]

(PS) 2 The area of a square is 49 cm². Work out the length of a side. [1]

Total Marks _____ / 4

(MR) 1 Jessa and Holly have been given the following question:

What is the value of $3 + 5 \times 4 + 7$?

Jessa thinks the answer is 30 and Holly thinks the answer is 39. Who is right? Explain your answer. [2]

(FS) 2 A netball club is planning a trip. The club has 354 members and the cost of the trip is £12 per member.

a) Work out the total cost of the trip. [3]

They need coaches for the trip and each coach seats 52 people.

b) How many coaches do they need to book? [3]

c) How many spare seats will there be? [2]

(MR) 3 Explain why $\sqrt{79}$ must be between 8 and 9 [2]

Total Marks _____ / 12

Sequences

1 **a)** Write down the next two numbers in the following sequence.

3, 7, 11, 15, _____ , _____,... [2]

b) Write down the term-to-term rule. [1]

2 Lynne plants a new flower in her garden.

Of the buds, five have already flowered.

Each week another three buds flower.

a) How many buds will have flowered after three weeks? [1]

b) How many weeks will it take for 32 buds to have flowered? [1]

Total Marks _____ / 5

1 **a)** Find the nth term of this arithmetic sequence.

4, 7, 10, 13, 16,... [3]

b) Find the 60th term in the sequence. [1]

(MR) **2** Match the cards on the left with the correct card on the right.

5, 9, 13, 17, 21,...	Neither
2, 8, 18, 32, 50,...	Quadratic
8, 17, 32, 53, 80,...	Arithmetic

[2]

Total Marks _____ / 6

Perimeter and Area 1

You must be able to:

- Find the perimeter and area of a square
- Find the perimeter and area of a rectangle
- Find the area of a triangle
- Find the area and perimeter of compound shapes.

Perimeter and Area of Squares and Rectangles

- The **perimeter** is the distance around the outside of a 2D shape.
- The formula for the perimeter of a square is $4 \times$ length or $P = 4l$
- The formula for the perimeter of a rectangle is:
  perimeter $= 2($length $+$ width$)$ or $P = 2(l + w)$
  also   perimeter $= 2($length$) + 2($width$)$ or $P = 2l + 2w$
- The formula for the **area** of a square or rectangle is:
  area $=$ length $\times$ width or $A = l \times w$

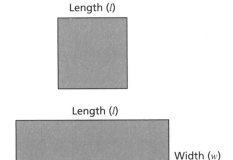

Length (*l*)

Length (*l*)

Width (*w*)

Find the perimeter and area of this square.

5 cm

5 cm

Perimeter $= 4 \times 5$
$= 20$ cm
Area $= 5 \times 5 = 25$ cm²

Find the perimeter and area of this rectangle.

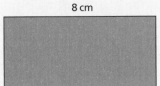

8 cm

3 cm

Perimeter $= 2(8 + 3)$
$= 2 \times 11$
$= 22$ cm
Area $= 8 \times 3$
$= 24$ cm²

Area of a Triangle

- The formula for the area of a triangle is:
  area $= \frac{1}{2}($base $\times$ **perpendicular** height$)$

Find the area of this triangle.

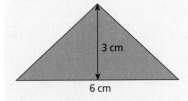

3 cm

6 cm

Area $= \frac{1}{2}(6 \times 3)$

$= \frac{1}{2}(18)$

$= 9$ cm²

> **Key Point**
>
> When finding the area of a triangle, always use the perpendicular height.

- The **altitude of a triangle** is a perpendicular line drawn from a vertex to the opposite side.

Area and Perimeter of Composite Shapes

- A **composite** shape is made up from other, simpler shapes.
- To find the area of a composite shape, divide it into basic shapes.

This shape can be broken up into three rectangles.

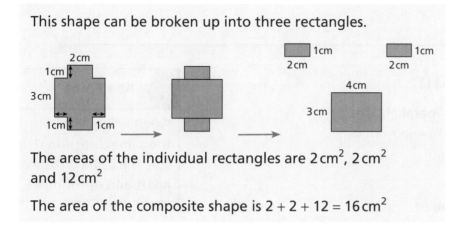

The areas of the individual rectangles are $2\,cm^2$, $2\,cm^2$ and $12\,cm^2$

The area of the composite shape is $2 + 2 + 12 = 16\,cm^2$

Key Point

Areas are two-dimensional and are measured in square units, for example cm^2

- To find the perimeter, start at one corner of the shape and travel around the outside, adding the lengths.

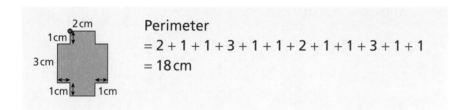

Perimeter
$= 2 + 1 + 1 + 3 + 1 + 1 + 2 + 1 + 1 + 3 + 1 + 1$
$= 18\,cm$

- Shapes that have straight edges and right angles are called **rectilinear** shapes.

Quick Test

1. Find the perimeter of a rectangle with width 5 cm and length 7 cm.
2. Find the area of a rectangle with width 9 cm and length 3 cm. Give appropriate units in your answer.
3. Find the area of a triangle with base 4 cm and perpendicular height 3 cm.
4. Find the perimeter and area of this shape.

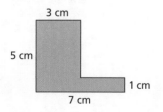

Key Words

perimeter
area
perpendicular
composite

Perimeter and Area 2

Quick Recall Quiz

You must be able to:

- Find the area of a parallelogram
- Find the area of a trapezium
- Find the circumference and area of a circle.

Area of a Parallelogram

- A **parallelogram** has two pairs of **parallel** sides.
- The formula for the area of a parallelogram is:

area = base × perpendicular height

Find the area of this parallelogram.

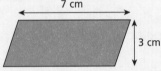

The base is 7 cm and the perpendicular height is 3 cm.

Area = 7 × 3

       = 21 cm^2

> **Key Point**
>
> Parallel lines travel in the same direction. They stay the same distance apart and never meet.

Area of a Trapezium

- A **trapezium** has one pair of parallel sides.
- The formula for the area of a trapezium is:

$A = \frac{1}{2}(a + b)h$

- The sides labelled a and b are the parallel sides and h is the perpendicular height.
- Perpendicular means at right angles.

Find the area of this trapezium.

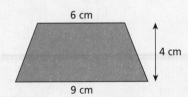

Area = $\frac{1}{2}(6 + 9) \times 4$

       = 30 cm^2

Circumference and Area of a Circle

- The **circumference** of a **circle** is the distance around the outside.
- The **radius** of a circle is the distance from the centre to the circumference.
- The **diameter** of a circle is twice the radius.
- The formula for the circumference of a circle is:
  $C = 2\pi r$ or $C = \pi d$
- The formula for the area of a circle is:
  $A = \pi r^2$

Find the circumference and area of this circle.
Give your answers to 1 decimal place.

The circumference:	The area:
$C = 2 \times \pi \times 7$	$A = \pi \times 7^2$
$= 14 \times \pi$	$= \pi \times 49$
$= 44.0\,\text{cm}$	$= 153.9\,\text{cm}^2$

- Circles can be split into **sectors**.
- A sector is a region bounded by two radii and an **arc** (a curved line that is part of the circumference).
- A sector with a 90° angle at the centre would have an area of $\frac{1}{4}$ of the whole circle.

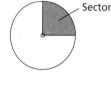

Sector

Find the perimeter and area of the sector of a circle.
Give your answers to 1 decimal place.

60° is one-sixth of 360°

Arc length $= \frac{1}{6} \times 2 \times \pi \times 4 = 4.2\,\text{cm}$

Perimeter $= 4.2 + 4 + 4 = 12.2\,\text{cm}$

Area $= \frac{1}{6} \times \pi \times 4^2 = 8.4\,\text{cm}^2$

Quick Test

1. Find the area of the parallelogram.

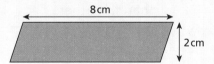

2. Find the area of the trapezium.

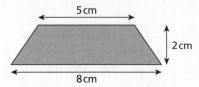

3. Find the circumference and area of a circle with radius 6 cm.
4. Find the circumference and area of a circle with diameter 4 cm.

Statistics

Statistics and Data 1

You must be able to:

- Find the mean, median, mode and range for a set of data
- Choose which average is the most appropriate to use in different situations
- Use a tally chart to collect data
- Construct bar charts and vertical line graphs.

Quick Recall Quiz

Mean, Median, Mode and Range

- The **mean** is the **sum** of all the values divided by the number of values.
- The **median** is the middle value when the data is in order.
- The **mode** is the most common value.
- The mean, median and mode are all averages and are called **measures of central tendency**.
- The **range** is the **difference** between the biggest and the smallest value.
- The range is a measure of spread or a **measure of dispersion**.

> **Key Point**
>
> Data can have more than one mode. Bi-modal means the data set has two modes.

Find the mean, median, mode and range for this data:
5, 9, 7, 6, 2, 7, 3, 11, 4, 2, 6, 6, 4

The mean $= \dfrac{5+9+7+6+2+7+3+11+4+2+6+6+4}{13} = 5.5...$

Mean = 5.5 (1 d.p.)

The median = 2, 2, 3, 4, 4, 5, ⑥ 6, 6, 7, 7, 9, 11

Median = 6

The mode is also 6 as this number is seen most often.

The range $= 11 - 2 = 9$

Put the data in order, smallest to biggest.

Choosing an Appropriate Average

- Use the **mode** when you are interested in the most common answer, for example if you were a shoe manufacturer deciding how many of each size to make.
- Use the **mean** when your data does not contain **outliers**. A company which wanted to find average sales across a year would want to use all values.
- Use the **median** when your data does contain outliers, for example finding the average salary for a company when the manager earns many times more than the other employees.

> **Key Point**
>
> An outlier is a value that is much higher or lower than the others.

Constructing a Tally Chart

- A **tally chart** is a quick way of recording data.
- Your data is placed into groups, which makes it easier to analyse.
- A tally chart can be used to make a frequency table by adding an extra column to record the total in each group.

> **Key Point**
>
> The frequency is the total for the group.

You are collecting and recording data about people's favourite flavour crisps. You ask 50 people and fill in the tally chart with their responses.

Flavour	Tally	Frequency
Plain	卌 卌 ||	12
Salt and vinegar	卌 ||||	9
Cheese and onion	卌 卌 卌 |	16
Prawn cocktail	卌 ||	7
Other	卌 |	6
Total		**50**

From the table, you can see that cheese and onion is the mode.

Bar Charts and Vertical Line Graphs

- **Bar charts** and **vertical line graphs** can be used to compare frequencies.

Draw a bar chart and a vertical line graph to represent the data above.

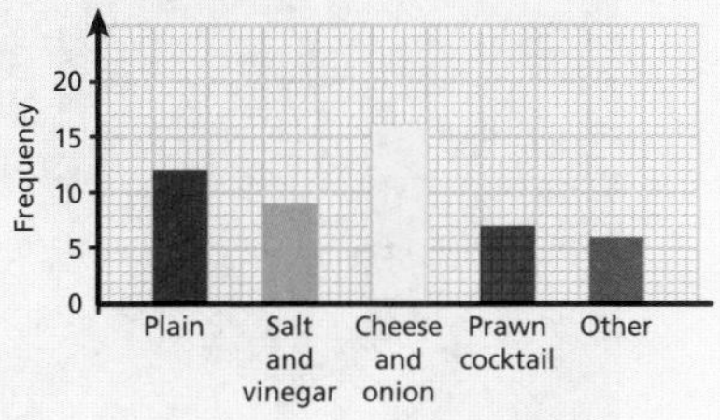

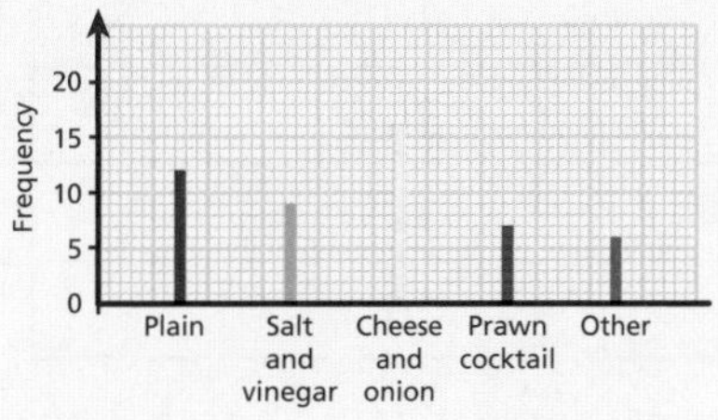

Quick Test

1. Emma surveyed her class to find out their favourite colour. She constructed the tally chart below.
   a) Complete the frequency column.
   b) How many students are there in Emma's class?

Colour	Tally	Frequency
Red	卌 ||||	
Blue	卌 ||	
Green	卌 |||	
Yellow	卌 |	
Other	卌 ||	

2. Look at this set of data: **3, 7, 4, 6, 3, 5, 9, 40, 6**
   a) Write down the mode of this data.
   b) Calculate the mean, median and range.
   c) Would you choose the mean or the median to represent this data? Explain your answer.

Key Words

mean
sum
median
mode
range
difference
outlier
tally chart
bar chart
vertical line graph

Statistics and Data 2

You must be able to:

- Group data and construct grouped frequency tables
- Construct and interpret a two-way table.

Grouping Data

- When you have a large amount of data, it is sometimes appropriate to place it into groups.
- A group is also called a **class interval**.
- The disadvantage of using **grouped data** is that the original **raw data** is lost.

> The data below represents the number of people who visited the library each day over a 60-day period.
>
> | Number of people | Frequency |
> |:---:|:---:|
> | 0–10 | 10 |
> | 11–20 | 30 |
> | 21–30 | 14 |
> | 31–40 | 6 |
>
> On 30 out of the 60 days, the library had between 11 and 20 (inclusive) visitors.

- Data such as the number of people is discrete as it can only take particular values.
- Data such as height and weight is continuous as it can take any value on a particular scale.
- This information can be shown using a bar chart.

> **Key Point**
>
> Calculations based on grouped data will be estimates.

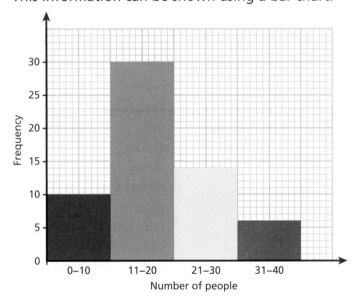

Two-Way Tables

- A two-way table shows information that relates to two different categories.
- Two-way tables can be constructed from information collected in a survey.

Daniel surveyed his class to find out if they owned any pets. In his class there are 16 boys and 18 girls. 10 of the boys owned a pet and 15 of the girls owned a pet.

	Pets	No Pets	Total
Boys	10		16
Girls	15		18
Total			

The information given is filled into the table and then the missing information can be worked out.

If there are 16 boys in Daniel's class and 10 of them have pets, then to work out how many boys do not have pets we calculate 16 – 10 = 6

When all the missing information is entered you can **interpret** the data.

	Pets	No Pets	Total
Boys	10	6	16
Girls	15	3	18
Total	25	9	34

You can see that there are 34 students in Daniel's class and 25 of them owned a pet. You can also see that more girls owned pets than boys, and that more pupils owned pets than did not own pets.

Quick Test

1. 25 women and 30 men were asked if they preferred football or rugby. 16 of the women said they prefer football and 10 of the men said they prefer rugby.
   a) Construct a two-way table to represent this information.
   b) How many in total said they prefer football?
   c) How many women preferred rugby?
   d) How many people took part in the survey?

Key Words

class interval
grouped data
raw data
interpret

Review Questions

Number

FS **1** I purchase a new car for £3000

I pay a deposit of £800 and pay the rest in four equal payments.

How much is each payment? [3]

2 The planets in our solar system orbit the Sun.
Their orbits are almost circular.
The table gives the time it takes, in days, for each
planet to complete one orbit of the Sun.

Earth	365
Jupiter	4332
Mars	687
Mercury	88
Neptune	60 200
Saturn	10 760
Uranus	30 700
Venus	224

Put the planets in order, starting with the shortest orbit time. [2]

3 A square has a perimeter of 20 cm.

Work out the area of the square. [2]

4 Work out $4 \times 2^2 + 6 \times 2^3$ [2]

Total Marks _____ / 9

MR **1** $48 \times 52 = 2496$

Use this to help you work out the following calculations.

$24 \times 52 =$ _____ $48 \times$ _____ $= 1248$ $2496 \div 52 =$ _____ [3]

PS **2** The lowest common multiple of two numbers is 60 and their sum is 27

What are the numbers? [2]

FS **3** Gautam is having a barbecue and wants to invite some friends.

Sausages come in packs of 6. Rolls come in packs of 8.

She needs exactly the same number of sausages and rolls.

What is the minimum number of each pack she can buy? [3]

Total Marks _____ / 8

Sequences

1 **a)** A function machine maps the number n to $n + 4$

Fill in the missing values.

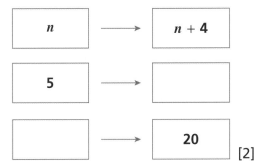

[2]

b) Many different function machines can map 36 to 6

Copy and complete the boxes below to give two **different** function machines.

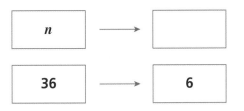 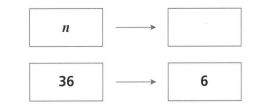

[2]

Total Marks .... / 4

1 An expression for the nth term of the arithmetic sequence 6, 8, 10, 12, … is $2n + 4$

a) Find the 20th term of this sequence. [1]

b) Find the 100th term of this sequence. [1]

c) The odd numbers also form an arithmetic sequence with a common difference of 2

Find the nth term for the sequence of odd numbers. [2]

PS **2** Lundy Island has two lighthouses; one in the north and one in the south.
The northern lighthouse flashes every 20 seconds and the southern lighthouse every 45 seconds.
Both lighthouses flash at 4 pm.

When do they next both flash at the same time? [3]

Total Marks .... / 7

Perimeter and Area

1 Work out the perimeter and area of the rectangle below.

6 cm

4 cm

[2]

(MR) **2** In the diagram below, the area of the triangle is 8 cm^2

Use this to help you work out the area of the rectangle.

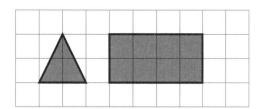

[2]

Total Marks .... / 4

(PS) **1** The area of this rectangle is 48 cm^2

Find the values of x and y.

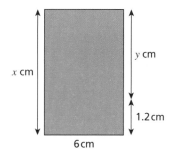

y cm

x cm

1.2 cm

6 cm

[2]

(FS) **2** Kelly is tiling her bathroom wall.

The wall is 4 m by 3 m.

Each tile is 25 cm by 25 cm.

a) Work out how many tiles Kelly needs to buy to tile the wall. [3]

The tiles come in packs of 10 and each pack costs £15

b) Work out how much it will cost Kelly to tile the wall. [2]

c) How many tiles will she have left over? [1]

Total Marks .... / 8

Statistics and Data

1 Look at these five numbers:

| 6 | 11 | 9 | 12 | 7 |

a) Show that the mean of the five numbers is 9 [1]

b) Explain why the median is also 9 [1]

c) Write down a different set of five numbers which also have a mean of 9 [1]

(PS) **2** The scores for a netball team are 15, 23, 8, 5, 19, x

The mean score is 15. What is x? [2]

3 The two-way table shows the number of bedrooms and the number of people living in each house in a street.

		Number living in house				
		1	2	3	4	5
Number of bedrooms	1	16	8	0	0	0
	2	5	12	18	3	0
	3	4	10	12	20	16
	4	2	6	7	15	11

a) How many houses have exactly one person living in them? [1]

b) How many houses have 3 bedrooms and exactly 4 people? [1]

c) How many houses have 4 bedrooms? [1]

Total Marks .... / 8

(MR) **1** Phil and Dave are both darts players. Their scores for a match are shown below.

| Phil | 64 | 70 | 80 | 100 | 57 | 100 | 41 | 56 | 30 |
| Dave | 36 | 180 | 21 | 180 | 10 | 5 | 23 | 25 | 140 |

a) Calculate the mean score for each player. [2]

b) Find the range of scores for each player. [2]

c) Only one of the two players can be picked to play in the next match.

Would you pick Phil or Dave? Explain your answer. [2]

Total Marks .... / 6

Decimals 1

You must be able to:

- Multiply and divide by 10, 100 and 1000
- Understand the powers of 10
- Understand standard form
- List decimals in size order.

Multiplying and Dividing by 10, 100 and 1000

- Multiplying by 10 moves the digits one place to the left.
- Multiplying by 100 moves the digits two places to the left.
- Multiplying by 1000 moves the digits three places to the left.

$$1.67 \times 10 = 16.7 \qquad 1.67 \times 100 = 167 \qquad 1.67 \times 1000 = 1670$$

- Dividing by 10 moves the digits one place to the right.
- Dividing by 100 moves the digits two places to the right.
- Dividing by 1000 moves the digits three places to the right.

$$360.7 \div 10 = 36.07 \quad 360.7 \div 100 = 3.607 \quad 360.7 \div 1000 = 0.3607$$

- You can use one calculation to work out the answer to another.

Use $32 \times 24 = 768$ to work out:
a) 320×240 **b)** 3.2×2.4 **c)** 16×24

a) $320 \times 240 = 32 \times 10 \times 24 \times 10$
$$= 768 \times 100 = 76\,800$$

b) $3.2 \times 2.4 = 32 \div 10 \times 24 \div 10$
$$= 768 \div 100 = 7.68$$

c) $16 \times 24 = \frac{1}{2} \times 32 \times 24$
$$= \frac{1}{2} \times 768 = 384$$

- A decimal that does **not** recur is called a **terminating decimal**.

Powers of Ten and Standard Form

- A **power** or **index** tells us how many times a number should be multiplied by itself.
- A power is also called an **exponent**.

$$10^2 = 10 \times 10 \qquad\qquad = 100 \qquad\qquad 10^{-1} = \frac{1}{10}$$

$$10^3 = 10 \times 10 \times 10 \qquad = 1000 \qquad\qquad 10^{-2} = \frac{1}{10^2} = \frac{1}{100}$$

$$10^4 = 10 \times 10 \times 10 \times 10 = 10\,000$$

> **Key Point**
>
> Multiplying by 10, 100 or 1000 makes the number bigger.

> **Key Point**
>
> Dividing by 10, 100 or 1000 makes the number smaller.

- **Standard form** or **standard index form** allows us to write very big and very small numbers more easily. Standard form uses powers of 10.
- A number **not** written in standard form is an **ordinary number**.
- Standard form is a number written in the form $A \times 10^n$, where $1 \leqslant A < 10$ and n is a positive or negative integer or zero.

> Write 2000 in standard form.
> $2000 = 2 \times 1000 = 2 \times 10^3$
>
> Write 2.5×10^4 as an ordinary number.
> $2.5 \times 10^4 = 2.5 \times 10\,000 = 25\,000$

> Put these numbers in order of size. Start with the smallest.
> 3.1×10^3 2.8×10^5 2.6×10^4 4.6×10^3
> Starting with the numbers with the lowest powers of 10, as $3.1 < 4.6$, the order is:
> 3.1×10^3 4.6×10^3 2.6×10^4 2.8×10^5

Ordering Decimals

- Place value can be used to compare decimal numbers.
- The digits after the decimal point are called tenths, hundredths, thousandths, and so on.

> Put in ascending order: 12.071, 12.24, 12.905, 12.902, 12.061

Each number starts with 12. So compare the tenths, hundredths and thousandths.

	Tens	Units	.	Tenths	Hundredths	Thousandths
12.071	1	2	.	0	7	1
12.24	1	2	.	2	4	0
12.905	1	2	.	9	0	5
12.902	1	2	.	9	0	2
12.061	1	2	.	0	6	1

12.071 and 12.061 are the two smallest as they have no tenths.

12.24 is the next smallest with 2 tenths.

12.905 and 12.902 are the two biggest as they have 9 tenths.

12.061 is smaller than 12.071 as it has only 6 hundredths compared to 7 hundredths.

12.902 is smaller than 12.905, as although they both have the same hundredths, 12.902 has only 2 thousandths compared to 5 thousandths.

So, in ascending order: 12.061, 12.071, 12.24, 12.902, 12.905

First group by the number of tenths.

Then order them within each group.

Key Point

Ascending order is smallest to biggest.

Descending order is biggest to smallest.

Key Words

terminating decimal
power
index
exponent
standard form
ordinary number

Quick Test

1. **a)** Work out 23.56×10 **b)** Work out $56.781 \div 10$
2. Write down the value of 10^5
3. Put the numbers in ascending order:
   16.34, 16.713, 16.705, 16.309, 16.2

Decimals 2

You must be able to:

- Add and subtract decimal numbers
- Multiply and divide decimal numbers
- Use rounding to estimate calculations.

Adding and Subtracting Decimals

- Decimal numbers can be added and subtracted in the same way as whole numbers.

> **Key Point**
>
> When adding and subtracting, line up the numbers by matching the decimal point.

Calculate 23.764 + 12.987

	2	3	.	7	6	4
+	1	2	.	9	8	7
	3	6 ^{1}	.	7 ^{1}	5 ^{1}	1

So 23.764 + 12.987 = 36.751

Calculate 12.697 − 8.2

	$\not{1}$	^{1}2	.	6	9	7
−	0	8	.	2	0	0
		4	.	4	9	7

So 12.697 − 8.2 = 4.497

Multiplying Decimals

- Complete the calculation without the decimal points and replace the decimal point at the end.
- Count how many digits are after the decimal points in the question and this is how many digits are after the decimal point in the answer.

Calculate 45.3 × 3.7
Here using the column method gives 453 × 37 = 16 761

		4	5	3	
	×		3	7	
	3	1 ^{3}	7 ^{2}	1	(7 × 453)
1	3 ^{1}	5	9	0	(30 × 453)
1	6	7 ^{1}	6	1	

There are two digits after the decimal points in the question.
So 45.3 × 3.7 = 167.61

Dividing Decimals

- **Equivalent** fractions can be used when dividing decimals.

Calculate $4.45 \div 0.05$

$$4.45 \div 0.05 = \frac{4.45}{0.05} = \frac{445}{5}$$

$$5\overline{)4\,^4 4\,^4 5}$$
$$8\quad9$$

So $4.45 \div 0.05 = 89$

Remember to multiply the numerator and denominator by the same amount.

Key Point

A division can be written as a fraction. Equivalent fractions are equal.

Rounding and Estimating

- Numbers can be **rounded** using **decimal places** (d.p.) or **significant figures** (s.f.).

Round 56.76 to 1 decimal place.

56.7|6

7 is the first decimal place and the digit after it is more than 5 so round 7 up to 8

56.76 to 1 decimal place is 56.8

- When **estimating** a calculation, round all the numbers to 1 s.f.

Key Point

The first significant figure is the first non-zero digit.

Estimate $26\,751 \times 64$

Round 2|6751

2 is the first significant figure. As the digit after it is 5 or more, round the 2 up to 3 and every digit after becomes zero.

So 26 751 to 1 significant figure is 30 000

Round 6|4

6 is the first significant figure. As the digit after it is less than 5, the 6 does not change and every digit after becomes zero.

So 64 to 1 significant figure is 60

An estimate for $26\,751 \times 64$ is $30\,000 \times 60 = 1\,800\,000$

- There is always an error to consider when a number is rounded.
- This error can be expressed using an inequality.
- If a number is rounded to 23 to the nearest whole number, the actual number could be anywhere between 22.5 and 23.5
  The rounding error can be expressed as $-0.5 \leqslant \text{error} < 0.5$

Key Words

equivalent
rounding
decimal places
significant figures
estimate

Quick Test

1. Work out $45.671 + 3.82$
2. Work out $34.321 - 17.11$
3. Work out $65.2 \div 0.4$
4. Estimate 3457×46

Algebra 1

You must be able to:

- Understand the vocabulary associated with algebra
- Know the difference between an equation and expression
- Collect like terms in an expression
- Write products as algebraic expressions
- Substitute numerical values into formulae and expressions.

Vocabulary in Algebra

- The numbers used in algebra are called **constants**.
- The letters used in algebra are called **variables**.
- A **term** is part of an **expression**, **equation** or **formula**.
- An expression is a collection of terms.
- An equation has an equals sign.
- A formula is a rule which links a variable to one or more other variables.

$2w + 3y + 6$ is an expression	$2w = 6$ is an equation	$P = 2w + 2l$ is a formula
$2w$ is a term	6 is a constant	w is a variable

- The number in front of a variable in a term is called the **coefficient**. For example, in the term $2w$ the coefficient is 2

Collecting Like Terms

- **Like terms** are terms with the same variables.
- To **simplify** an expression, like terms are collected.

Simplify $5x + 8y + 3x - y$

Collect the like terms: $5x + 3x + 8y - y$

The x terms can be simplified: $5x + 3x = 8x$

The y terms can be simplified: $8y - y = 7y$

So $5x + 8y + 3x - y$ can be simplified to $8x + 7y$

> In this example there are x terms and y terms.

Key Point

Remember that terms have a + or − sign between them and each sign belongs to the term on its right.

Simplify $2x^2 + 6y - x^2 + 4y - 6$

Collect the like terms: $2x^2 - x^2 + 6y + 4y - 6$

The x^2 terms can be simplified: $2x^2 - x^2 = x^2$

The y terms can be simplified: $6y + 4y = 10y$

There is only one constant term: -6

So $2x^2 + 6y - x^2 + 4y - 6$ can be simplified to $x^2 + 10y - 6$

> In this example, there are x^2 terms, y terms and constant terms.

Simplify $\frac{2}{3}x + y - \frac{1}{3}x + \frac{3}{4}y$

Collect the like terms: $\frac{2}{3}x - \frac{1}{3}x + y + \frac{3}{4}y$

The x terms can be simplified: $\frac{2}{3}x - \frac{1}{3}x = \frac{1}{3}x$

The y terms can be simplified: $y + \frac{3}{4}y = \frac{7}{4}y$

So $\frac{2}{3}x + y - \frac{1}{3}x + \frac{3}{4}y$ can be simplified to $\frac{1}{3}x + \frac{7}{4}y$

In this example, some of the coefficients are fractions.

Expressions with Products

- The product of a and b is the same as $a \times b$ or ab
- Expressions with products are usually written without the $\times$ sign.

$2 \times a = 2a$

$a \times b = ab$

$a \times a = a^2$

$a \times a \times b = a^2b$

$a \times a \times a = a^3$

$a \div (b \times c) = \frac{a}{bc}$

> ### Key Point
>
> There are many scientific problems which involve substituting into formulae. Always follow BIDMAS.
>
> To find the perimeter, area and volume of shapes, we substitute into a formula.

Substitution

- Variables in formulae can be written in shorthand by representing them with a letter.
- Some commonly used scientific formulae are:

speed $= \dfrac{\text{distance}}{\text{time}}$ in shorthand $s = \dfrac{d}{t}$

density $= \dfrac{\text{mass}}{\text{volume}}$ in shorthand $d = \dfrac{m}{v}$

- Substitution involves replacing the letters in a given formula or expression with numbers.

Find the value of the expression $2a + b$ when $a = 3$ and $b = 5$

Replace the letters with the given values.

$2a + b = 2 \times 3 + 5 = 11$

$\frac{3}{4} = 0.75$

Emma took $\frac{3}{4}$ of an hour to travel 30 miles.

Calculate her average speed in miles per hour.

$s = \dfrac{d}{t}$

$s = \dfrac{30}{0.75} = 40\,\text{mph}$

> ### Key Words
>
> constant
> variable
> term
> expression
> equation
> formula
> coefficient
> like terms
> simplify

> ### Quick Test
>
> 1. Simplify $4x + 7y + 3x - 2y + 6$
> 2. Simplify $c \times c \times d \times d$
> 3. Find the value of $4x + 2y$ when $x = 2$ and $y = 3$

Algebra 2

You must be able to:

- Understand the commutative, associative and distributive laws
- Multiply a single term over a bracket
- Carry out binomial expansion
- Factorise linear expressions
- Change the subject of a formula.

Fundamental Laws

- The **commutative law** says that $a + b = b + a$, e.g. $2 + 3 = 3 + 2$
- The **associative law** says that $(a + b) + c = a + (b + c)$, e.g. $(2 + 3) + 4 = 2 + (3 + 4)$
- The **distributive law** says that $a(b + c) = ab + ac$, e.g. $2(3 + 4) = 2 \times 3 + 2 \times 4$

Work out the answers to check these are true.

> **Key Point**
>
> A reminder of the rules when multiplying and dividing negative numbers:

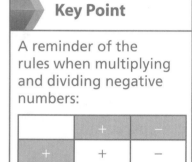

	+	−
+	+	−
−	−	+

Expanding Brackets

- **Expanding** involves removing the brackets by **multiplying every term** inside the bracket by the number or term on the outside.

Expand $3(x + 5)$

×	3
x	$3x$
$+5$	$+15$

This is the distributive law.

$3(x + 5) = 3x + 15$

Expand and simplify $4(x + y) - 2(2x - 3y)$

×	4
x	$4x$
y	$4y$

$= 4x + 4y$

×	2
$2x$	$4x$
$-3y$	$-6y$

$= 4x - 6y$

Then collect like terms: $(4x + 4y) - (4x - 6y) = 10y$

Binomial Expansion

- A **binomial** is an expression that contains two terms, e.g. $x + 3$ or $3y - 1$
- The product of two (or more) binomials is when they are multiplied together, e.g. $(x + 4)(3x - 1)$
- To expand (or multiply out) the brackets, every term in the first bracket is multiplied by every term in the second bracket.

Expand and simplify $(x + 3)(x + 5)$

×	x	$+3$
x	x^2	$+3x$
$+5$	$+5x$	15

$(x + 3)(x + 5) = x^2 + 3x + 5x + 15$
$= x^2 + 8x + 15$

Simplify by collecting like terms.

Expand and simplify $(3y - 2)(2y + 7)$

×	$3y$	-2
$2y$	$6y^2$	$-4y$
$+7$	$+21y$	-14

$(3y - 2)(2y + 7) = 6y^2 - 4y + 21y - 14$
$= 6y^2 + 17y - 14$

Simplify by collecting like terms.

Factorising

- **Factorising** (or factorisation) is the reverse of expanding brackets.
- To factorise, look for **common factors**.

Factorise $6x + 9$

×	3
	6x
	+9

×	3
2x	6x
+3	+9

So $6x + 9 = 3(2x + 3)$

3 is a common factor of 6 and 9 so take the 3 to the outside of the bracket. To find what is inside the bracket, fill in the blanks in the table.

Factorise $x^2 + 2x$

x is the common factor:

So $x^2 + 2x = x(x + 2)$

×	x
x	x²
+2	+2x

Factorise $6x^3 + 2x^2$

2 and x^2 are the common factors:

So $6x^3 + 2x^2 = 2x^2(3x + 1)$

×	2x²
3x	6x³
+1	2x²

Changing the Subject of a Formula

- The **subject** of a formula appears once on the left-hand side.
- To change the subject, a formula must be rearranged using **inverse** operations. Write the new subject on the left-hand side of the rearranged formula.

Make x the subject of $y = 3x + 1$

$y - 1 = 3x$ ← Subtracting 1 from both sides.

$\dfrac{y - 1}{3} = x$ or $x = \dfrac{y - 1}{3}$ ← Dividing both sides by 3

Make r the subject of $A = \pi r^2$ ← This is the formula for the area of a circle with radius r.

$\dfrac{A}{\pi} = r^2$ ← Dividing both sides by π

$\sqrt{\dfrac{A}{\pi}} = r$ or $r = \sqrt{\dfrac{A}{\pi}}$ ← Taking the square root of both sides.

This is the formula for the radius of a circle with area A.

Quick Test

1. Expand $4(2x - 1)$
2. Expand and simplify $2(2x - y) - 2(x + 6y)$
3. Factorise $5x - 25$
4. Factorise completely $2x^2 - 4x$

Review Questions

Perimeter and Area

1 Find the area of a circle with radius 7 cm. [2]

2 Find the perimeter and area of the rectangle below.

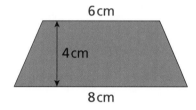
5 cm

9 cm

[2]

3 Find the area of the trapezium below.

6 cm

4 cm

8 cm

[2]

Total Marks _____ / 6

1 Frances wants to paint the side of her house.
The diagram represents the side of her house.

a) Find the area that Frances needs to paint. [3]

Each tin of paint covers 7 m²

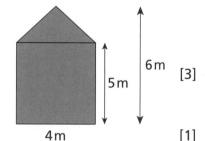
6 m
5 m
4 m

b) Work out the number of tins that Frances needs to buy. [1]

Each tin costs £12

c) Work out how much it will cost Frances to paint the side of her house. [2]

(PS) **2** Edmund buys a new bicycle and he uses it to cycle 8000 m to work every day.

His wheel is a circle with a diameter of 50 cm.

Work out how many times his wheel makes a full rotation during his journey. [2]

Total Marks _____ / 8

Statistics and Data

1 There are 34 pupils in class 8D and 19 of those are boys.

26 of the class are right-handed and 14 of those are girls.

Use this information to complete the table below. [3]

	Boys	Girls
Right-handed		
Left-handed		

2 James does a survey of the favourite vegetables of the pupils in his class:

carrots	*potatoes*	*peas*	*sweetcorn*	*peas*	*peas*
potatoes	*peas*	*carrots*	*peas*	*potatoes*	*sweetcorn*
potatoes	*potatoes*	*potatoes*	*carrots*	*carrots*	*potatoes*
peas	*potatoes*	*sweetcorn*	*peas*	*potatoes*	*peas*

a) Construct a frequency table to represent this data. [2]

b) What was the mode? [1]

Total Marks _____ / 6

 1 This data below shows the number of people attending six matches for Sandex United.

| 1240 | 1354 | 1306 | 14808 | 1378 | 1430 |

You want to calculate the average attendance.

a) Would you find the mean, median or mode? Give a reason for your answer. [2]

b) Which of the attendance numbers is an outlier? [1]

Total Marks _____ / 3

Practice Questions

Decimals

1 Match the card on the left with the correct card on the right.

One has been done for you.

3.7 × 10	3700
3.7 × 100	0.037
3.7 ÷ 10	370
3.7 ÷ 100	37
3.7 × 1000	0.37

[3]

2 Which is greater, 0.5679 or 0.5588?

Explain your answer. [2]

Total Marks .... / 5

1 Work out the following:

a) 34.542 + 23.29 [1]

b) 65.21 − 43.23 [1]

c) 21.81 × 3.4 [1]

d) 43.2 ÷ 0.2 [1]

2 Put these decimals in order of size. Start with the smallest.

17.84, 17.203, 17.09, 16.94, 16.061 [2]

Total Marks .... / 6

Algebra

1 Complete the statements.

When $x = 5$ $5x =$ _____

When $x =$ _____ $5x = 45$

When $x = 7$ _____ $= 28$ [2]

2 In this algebra grid, the total on each brick is made from the sum of the two bricks below. The first brick has been calculated for you. Copy and complete the grid.

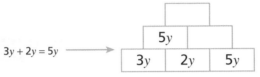

$3y + 2y = 5y$

| | 5y | |
| 3y | 2y | 5y |

[2]

Total Marks _____ / 4

(MR) **1** Claudia thinks the perimeter of this rectangle is $2x + 2y$

Lawrence thinks the perimeter of the rectangle is $2(x + y)$

Who is right – Claudia, Lawrence or both of them?
Explain your answer.   [2]

(FS) **2** The cost (in £) of hiring a car for the day and driving it y miles is given by this formula:

$C = 75 + 0.4y$

Work out how much it would cost to hire the car for a day and travel 120 miles.   [3]

3 Expand and simplify $2(4x + 1) - 5(x - 1)$ [2]

4 Factorise completely $3abc + 6a$ [2]

(PS) **5** If $ab = 36$ and $a = 4$, find the value of a^2b [2]

6 Copy and complete the algebra grid. The total in each brick is made from the sum of the two bricks below.

	5a − b	
	2a	
a		

[2]

Total Marks _____ / 13

3D Shapes: Volume and Surface Area 1

Quick Recall Quiz

You must be able to:

- Name and draw 3D shapes
- Draw the net of a cuboid and other 3D shapes
- Calculate the surface area and volume of a cuboid.

Naming and Drawing 3D Shapes

- A 3D shape can be described using the number of **faces**, **vertices** and **edges** it has.
- A **prism** is a solid shape with a uniform **cross-section**. The cross-section of a prism can be any polygon.
- Cubes and cuboids are prisms.

Shape	Name	Edges	Vertices	Faces
	Cube	12	8	6
	Cuboid	12	8	6
	Triangular prism	9	6	5
	Square-based pyramid	8	5	5
	Cylinder	2	0	3
	Pentagonal prism	15	10	7
	Cone	1	1	2
	Sphere	0	0	1

> **Key Point**
>
> A face is a sur'face', for example a flat side of a cube.
>
> A vertex is where edges meet, for example the corner of a cube.
>
> An edge is where two faces join.

Nets for 3D Shapes

- To create the **net** of a cuboid, imagine it is a box you are unfolding to lay out flat.

- Nets for a triangular prism and a square-based pyramid look like this:

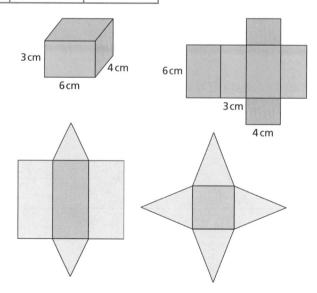

Surface Area of a Cube or Cuboid

- The **surface area** of a cube or cuboid is the sum of the areas of all six faces. The units for area are cm^2, m^2, etc. On a cube all the faces have the same area.

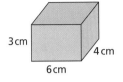

To calculate the surface area of the cuboid shown right, you can use the net to help you. Work out the area of each rectangle by multiplying the base by its height:

Green rectangle: 6 cm × 4 cm = 24 cm^2
There are two of them, so 24 cm^2 × 2 = 48 cm^2

Blue rectangle: 3 cm × 6 cm = 18 cm^2
There are two of them, so 18 cm^2 × 2 = 36 cm^2

Pink rectangle: 4 cm × 3 cm = 12 cm^2
There are two of them, so 12 cm^2 × 2 = 24 cm^2

Sum of all six areas is 48 + 36 + 24 = 108 cm^2

Volume of a Cuboid

- **Volume** is the space contained inside a 3D shape.

To calculate the volume of the cuboid shown right, you need to first work out the area of the front rectangle:

3 cm × 6 cm = 18 cm^2

Next multiply this area by the depth of the cuboid, 4 cm.

18 cm^2 × 4 cm = 72 cm^3

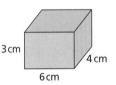

← The units are cm^3 this time.

- You can use the formula Volume of a cuboid (V) = Length (l) × width (w) × height (h) or $V = lwh$

Find the volume of a cuboid measuring 6 cm by 3 cm by 2 cm.

$V = 6 × 3 × 2$
$ = 36 \text{ cm}^3$

Quick Test

Work out the volume and the surface area of these cuboids.

1.

2.

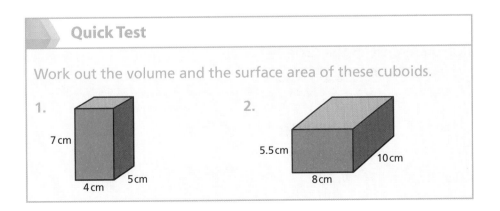

Key Words

face
vertex
edge
prism
cross-section
net
surface area
volume

3D Shapes:
Volume and Surface Area 2

Quick Recall Quiz

You must be able to:

- Calculate the volume and surface area of a cylinder
- Calculate the volume and surface area of a prism
- Calculate the volume of composite shapes.

Volume and Surface Area of a Cylinder

- You work out the volume of a **cylinder** the same way as the volume of a cuboid. First work out the area of the circle and then multiply it by the height of the cylinder.

> Calculate the volume of this cylinder.
> Volume = $(\pi \times 5 \times 5) \times 3$
> $= 235.62 \text{ cm}^3$ (2 d.p.)

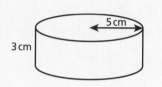

> **Key Point**
>
> Diameter is the full width of a circle that goes through the centre.
>
> Radius is half of the diameter.

← Area of a circle = $\pi \times \text{radius}^2$

> Calculate the volume of this cylinder.
>
> The cylinder has a diameter of 12 cm.
>
> This means the radius is 6 cm.
>
> Volume = $(\pi \times 6 \times 6) \times 5$
> $= 565.49 \text{ cm}^3$ (2 d.p.)

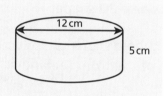

← Volume units are shown by a ^{3}, for example cm^3.

- To calculate the surface area of a cylinder, first draw the net (imagine cutting a can open).

> Calculate the surface area of this cylinder.

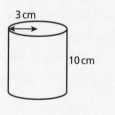
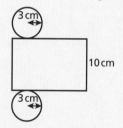

> The circumference of the circular end is the same as the length of the rectangle.
>
> Calculate the area of the circular ends $(\pi \times 3 \times 3) \times 2 = 56.548\ldots$
>
> The area of the rectangle is the circumference of the circle multiplied by the height of the cylinder, 10 cm.
>
> $(\pi \times 6) \times 10 = 188.495\ldots$
>
> Add the areas together: $56.548\ldots + 188.495\ldots = 245.04 \text{ cm}^2$ (2 d.p.)

> **Key Point**
>
> Circumference of a circle is the perimeter.

← Because there are two circles.

← Circumference = $\pi \times$ diameter
Diameter is double the radius.

- You can use the formula
  Curved surface area of a cylinder $C = 2\pi rh$ or $C = \pi dh$

Volume and Surface Area of a Prism

- Volume of prism = Area of cross-section × length or $V = Al$
- The surface area of a prism is the sum of the areas of all the faces.

Work out the volume and surface area of the triangular prism.

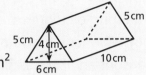

Area of cross-section is $\frac{1}{2} \times 6 \times 4 = 12\,cm^2$

Volume of prism is $12 \times 10 = 120\,cm^3$

Surface area =

2 × area of cross-section + area of the three rectangles

$$= 12 + 12 + (10 \times 6) + (10 \times 5) + (10 \times 5)$$
$$= 12 + 12 + 60 + 50 + 50$$
$$= 184\,cm^2$$

Volume of Composite Shapes

- **Composite** means the shape has been 'built' from more than one shape.

This shape is built from two cuboids.

Calculate the volume of the two separate cuboids and add the volumes together.

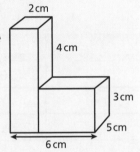

$(6\,cm - 2\,cm) \times 5 \times 3 = 60\,cm^3$

$(3\,cm + 4\,cm) \times 2 \times 5 = 70\,cm^3$

$60 + 70 = 130\,cm^3$

> **Key Point**
>
> Use the shape's dimensions to work out missing lengths.

Quick Test

1. Work out the volume and the surface area of these shapes.

   a)
   b)

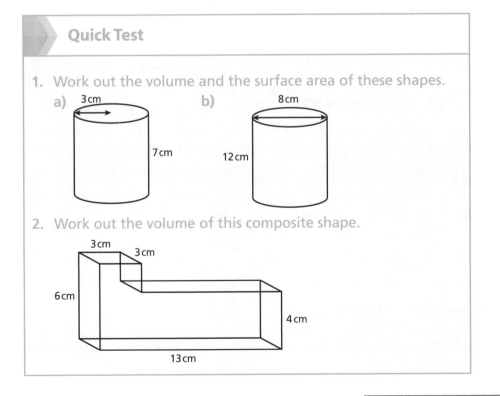

2. Work out the volume of this composite shape.

> **Key Word**
>
> cylinder

Interpreting Data 1

Quick Recall Quiz

You must be able to:

- Create a simple pie chart from a set of data
- Create and interpret pictograms
- Use frequency tables and draw frequency diagrams
- Make comparisons and contrasts between data.

Pie Charts

- **Pie charts** are often shown with **percentages** or **angles** indicating sector size – this and the visual representation helps to interpret the **data**.

36 students were asked the following question:
Which is your favourite flavour of crisps?
To work out the angle for
each person: 360 ÷ 36 = 10°

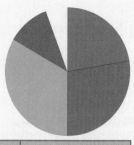

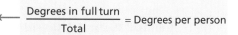

$$\frac{\text{Degrees in full turn}}{\text{Total}} = \text{Degrees per person}$$

Flavour of crisps	No. of students	Degrees
Salt and Vinegar ■	8	8 × 10 = 80
Cheese and Onion ■	10	10 × 10 = 100
Ready Salted ■	12	12 × 10 = 120
Prawn Cocktail ■	4	4 × 10 = 40
Other □	2	2 × 10 = 20

> **Key Point**
>
> Always align your protractor's zero line with your starting line.
>
> Count up from zero to measure your angle.
>
> Don't forget to label your chart.

Pictograms

- Data is represented by a picture or symbol in a **pictogram**.

The pictogram shows how many pizzas were delivered by Ben in one week. Key = 8 pizzas

Day	Mon	Tue	Wed	Thu	Fri	Sat	Sun
Pizza deliveries	●● ●	●	●◗	●	●● ●●	●●	

How many pizzas did Ben deliver on Friday? 8 × 4 = 32

On Sunday Ben delivered 20 pizzas. Complete the pictogram.

20 ÷ 8 = 2.5

Frequency Tables and Frequency Diagrams

- **Frequency tables** can be used to collect data into groups.
- **Frequency diagrams** (or frequency charts) illustrate the information.

The frequency table shows the heights of 100 students.

Height (h cm)	Frequency
$70 < h \leqslant 80$	3
$80 < h \leqslant 90$	4
$90 < h \leqslant 100$	12
$100 < h \leqslant 110$	24
$110 < h \leqslant 120$	30
$120 < h \leqslant 130$	22
$130 < h \leqslant 140$	3
$140 < h \leqslant 150$	2

Using the span of each height category, plot each group as a block using the frequency **axis**.

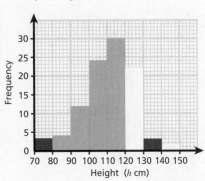

Data Comparison

- You can make comparisons using frequency diagrams.

These graphs show how much time four students spend on their mobile phones in a week. Compare the two graphs.

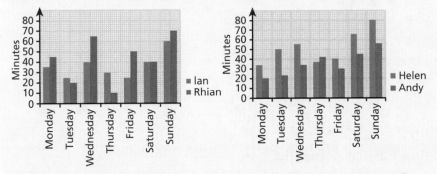

Apart from Thursday, Helen uses her phone more than Andy. In Ian's and Rhian's data, Rhian uses her phone more overall. Comparing both sets of data, we can say the girls use their phones more than the boys.

Key Point

Think about what is similar about the data and what is different. Look for any patterns.

Key Words

pie chart
percentage
angle
data
pictogram
frequency table
frequency diagram
axis

Quick Test

1. If 18 people were asked a question and you were to create a pie chart to present your data, what angle would represent one person?
2. Using the graph above right, on which day did both Helen and Andy use their mobiles the most?
3. The following week, Andy had a mean use of 40 minutes per day. Is this greater than Helen's use the previous week?

Interpreting Data 2

You must be able to:

- Interpret different graphs and diagrams
- Draw a scatter graph and understand correlation
- Understand the use of statistical investigations.

Interpreting Graphs and Diagrams

- You can interpret the information in graphs and diagrams.

What does this graph show?

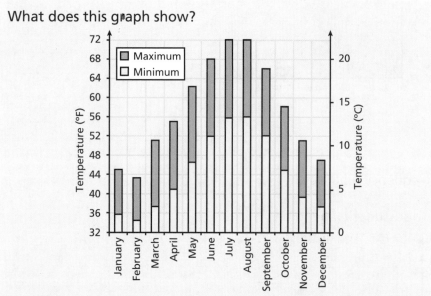

Using the labels and key, you can see that it gives the temperature and months of a year, plus a maximum and minimum temperature.

Which month has the lowest temperature?

Using the minimum temperature (yellow bars), pick the smallest bar, in this case, February.

Look at this pie chart. What is the most likely way the team scores a goal?

The largest sector of the pie chart represents scoring a goal in free play.

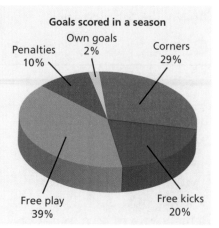

Goals scored in a season

Own goals 2%

Penalties 10%

Corners 29%

Free play 39%

Free kicks 20%

Drawing a Scatter Graph

- A **scatter graph** is used to investigate a relationship between two variables.
- Data for two variables is called **bivariate data**.
- If a linear relationship exists, a **line of best fit** can be drawn through the points on a scatter graph.

> Here is a plot of the sales of ice cream against the amount of sun per day for 12 days. The scatter graph shows that when the weather is sunnier (and hotter), more ice creams are sold.
>
>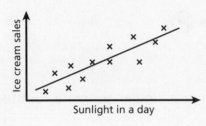

- You can describe the **correlation** and use the line of best fit to estimate data values.
- The graph above shows a positive correlation between ice cream sales and sunlight in a day.

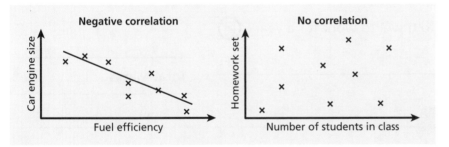

Statistical Investigations

- Statistical investigations use **surveys** and experiments to test statements and theories to see whether they might be true or false. These statements are called **hypotheses**.
- Survey questions should be specific, timely and have no overlap on answer choices.

Key Point

Plot each data pair as a coordinate. A line of best fit doesn't have to start at zero.

This scatter graph is showing bivariate data.

Key Point

Surveys should be made on a large random sample, never just a limited few.

Key Words

scatter graph
line of best fit
correlation
survey
hypothesis

Quick Test

1. Name two things you might plot against one another to show a positive correlation.
2. Look at the pie chart on page 48. What is the least likely way that the team scored a goal?

Decimals

1 Work out 2.53×0.3  [2]

2 Work out $0.85 \div 0.05$ [2]

3 Circle the two numbers which add to make 9 [1]

  8.1   **0.7**   **5.2**   **3.8**   **0.8**

4 By rounding both numbers to 1 significant figure, find an estimate for the following:

$\dfrac{6782}{53}$ [2]

5 I pay £17.80 to travel to work each week.

I work for 48 weeks a year.

How much does it cost me to travel to work for a year? [2]

Total Marks _____ / 9

---

1 Put the following numbers in order, from smallest to biggest.

7.765, 7.675, 6.765, 7.756, 6.776 [2]

2 Thomas buys three books that cost £2.98, £3.47 and £9.54

a) How much did the books cost in total? [2]

b) How much change should he get from a £20 note? [1]

3 Use your calculator to work out $\sqrt{48}$ to 2 decimal places. [1]

4 A number has been rounded to 35 to the nearest whole number.

Express the rounding error as an inequality. [1]

Total Marks _____ / 7

Algebra

PS **1** Simplify these expressions:

    a) $5k + 5 + 6k$ **b)** $k + 2 + 3k - 1$

    [2]

2 Fill in the missing term in the statement below.

    $3k + 4$ .... $= k + 4$

    [1]

3 Copy the table and complete the missing information. The first row has been done for you.

$c \times d$	cd
$c \times d \times d$	
$c \times c \times d$	
$c \times c \times d \times d$	

    [2]

Total Marks .... / 5

MR **1** This rectangle has dimensions $a \times b$.

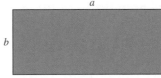

    a) Write a simplified expression for the area of the rectangle.   [1]

    b) Write a simplified expression for the perimeter of the rectangle.   [1]

    c) Another rectangle has an area of $15a^2$ and a perimeter of $16a$.
        What are the dimensions of this rectangle?   [1]

2 Factorise the following expression completely.

    $8ut^2 - 4ut + 20t$ [2]

3 Lucy walked 1200 metres in 20 minutes.

    $s = \dfrac{d}{t}$

    Use the formula to find her average speed in metres per hour.   [2]

Total Marks .... / 7

3D Shapes: Volume and Surface Area

1 Copy and complete the table below.

Shape	Name	No. of faces	No. of edges	No. of vertices

[3]

2 Name the 3D solid for each net.

a)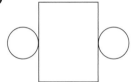

b)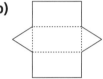

c)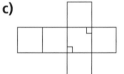

[3]

3 Calculate the volume and surface area of these two cuboids.

a)

10 cm
3 cm
5 cm

b)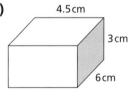

4.5 cm
3 cm
6 cm

[4]

Total Marks _____ / 10

(MR) **1** Find the height of a cylinder with a radius of 5 cm and a volume of 942 cm^3
Give your answer to 1 decimal place. [2]

(MR) **2** Find the height of a cylinder with a radius of 7 cm and a volume of 1385 cm^3
Give your answer to 2 decimal places. [2]

Total Marks _____ / 4

Interpreting Data

(MR) **1** 45 people were asked what their favourite cheese was.

Copy and complete the table and plot the information on a pie chart.

Category	Frequency	Angle
Brie	21	
Cheddar	5	
Stilton	14	
Other	5	
Total		

[6]

(MR) **2** One hundred students take two maths tests. The results are shown on the graph.

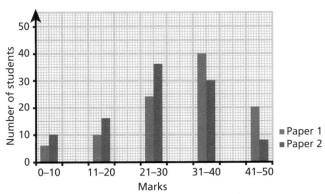

How many more students had marks 41–50 on paper 1 than on paper 2? [1]

Total Marks _____ / 7

1 What two things could you plot against one another to show a negative correlation? [2]

2 Write a question with response box options to determine whether people shop more over the Christmas period than at other times of the year. [4]

(MR) **3** a) Name four different types of statistical graphs or charts. [2]

b) Which one would you use to plot the information collected from asking 40 students, 'How long do you spend doing homework in a week?' Give a reason. [2]

Total Marks _____ / 10

 **Number**

Fractions 1

You must be able to:

- Find equivalent fractions
- Order fractions
- Add and subtract fractions.

Equivalent Fractions

- Equivalent fractions are fractions that are equal despite the **denominators** being different.

$\frac{1}{2}$ $\frac{2}{4} = \frac{1}{2}$ $\frac{4}{8} = \frac{1}{2}$

- You can create an equivalent fraction by keeping the ratio between the **numerator** and denominator the same.
- You do this by multiplying or dividing both the numerator and denominator by the same number.
- Creating equivalent fractions is very useful when you want to compare or evaluate different fractions.

> **Key Point**
>
> The numerator is the top part of a fraction.
>
> The denominator is the bottom part of a fraction.

Ordering Fractions

- You can use equivalent fractions to compare the size of fractions.

Which is the larger fraction, $\frac{2}{5}$ or $\frac{3}{7}$?

$\frac{2}{5}$ $\frac{3}{7}$

To compare these fractions, you need to find a common denominator – a number that appears in both the 5 and 7 times tables.

$5 \times 7 = 35$

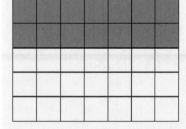 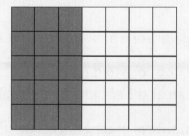

So $\frac{2}{5}$ becomes $\frac{14}{35}$ and $\frac{3}{7}$ becomes $\frac{15}{35}$

Now the denominators are equal, you can compare the two fractions more easily and you can see that $\frac{15}{35} = \frac{3}{7}$ is larger.
So $\frac{3}{7} > \frac{2}{5}$

> **Key Point**
>
> A common denominator is a number that shares a relationship with both fractions' denominators.
>
> For example, for 5 and 3 this would be 15, 30, 45, 60, …

Adding and Subtracting Fractions

- Adding fractions with the same denominator is straightforward. The numerators are collected together.

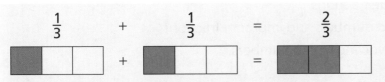

Notice the size of the 'piece', the denominator, remains the same in both the question and the answer.

- When subtracting fractions with the same denominator, simply subtract one numerator from the other.

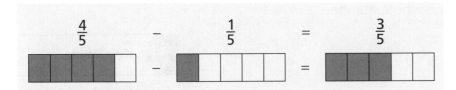

Key Point

The size of the 'piece' (the denominator) has to be the same to perform either addition or subtraction.

- When you have fractions with different denominators, first find equivalent fractions with a common denominator.

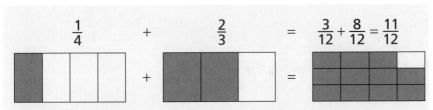

Here the common denominator is 12, as it is the smallest number that appears in both the 3 and 4 times tables.

This means that, for the first fraction, you have to multiply both the numerator and denominator by 3 and for the second fraction multiply them by 4

Now the fractions are of the same size 'pieces', you can add the numerators as before.

Key Point

It is essential to find equivalent fractions so both fractions have the same denominator.

Quick Test

1. Find three equivalent fractions for $\frac{2}{3}$

2. Work out $\frac{2}{7} + \frac{6}{11}$

3. Work out $\frac{7}{9} - \frac{3}{8}$

4. Work out $\frac{7}{13} - \frac{1}{4}$

5. Work out $\frac{14}{25} + \frac{3}{5} - \frac{7}{20}$

Key Words

denominator
numerator

Fractions 2

You must be able to:

- Multiply and divide fractions
- Understand mixed numbers and improper fractions
- Calculate sums involving mixed numbers.

Multiplying and Dividing Fractions

- Multiplying fractions by whole numbers is not very different from multiplying whole numbers.
- The numerator is multiplied by the whole number.

$$\frac{2}{7} \times 3 = \frac{2}{7} + \frac{2}{7} + \frac{2}{7} = \frac{6}{7}$$

> **Key Point**
>
> No common denominator is needed for multiplying or dividing fractions.

- When multiplying, you multiply the numerators and then multiply the denominators.

$$\frac{3}{5} \times \frac{2}{7} = \frac{3 \times 2}{5 \times 7} = \frac{6}{35}$$

When you multiply these two fractions, it is like saying there are $\frac{3}{5}$ lots of $\frac{2}{7}$

- When dividing fractions, use **inverse** operations. You change the operation to a multiplication and invert the second fraction.

$$\frac{2}{5} \div \frac{1}{2} = \frac{2 \times 2}{5 \times 1} = \frac{4}{5}$$

$\frac{2}{5}$ divided into halves gives twice as many pieces.

Mixed Numbers and Improper Fractions

- A **mixed number** is where there is both a whole number part and a fraction, for example $1\frac{1}{3}$
- An **improper fraction** is where the numerator is bigger than the denominator, for example $\frac{4}{3}$

Change the improper fraction $\frac{14}{3}$ to a mixed number.

$$\frac{14}{3} = 4\frac{2}{3}$$

How many 3s are there in 14?

The remainder is left as a fraction.

Change the mixed number $5\frac{3}{4}$ to an improper fraction.

$$5\frac{3}{4} = \frac{(5 \times 4) + 3}{4} = \frac{23}{4}$$

Multiply the whole number by the denominator. 5 units is 20 quarters.

Add the $\frac{3}{4}$ to make 23 quarters.

Key Point

Improper fractions can be changed into a mixed number and vice versa.

Adding and Subtracting Mixed Numbers

- To add and subtract with mixed numbers, you first have to change them to improper fractions.
- There are four steps for addition:
  convert any mixed numbers to improper fractions → convert to equivalent fractions with a common denominator → add → convert back to a mixed number

$$2\frac{4}{9} + 3\frac{1}{4} = \frac{22}{9} + \frac{13}{4} = \frac{22 \times 4}{36} + \frac{13 \times 9}{36} = \frac{205}{36} = 5\frac{25}{36}$$

- The same sequence of steps is needed for subtraction:
  convert any mixed numbers to improper fractions → convert to equivalent fractions with a common denominator → subtract → convert back to a mixed number

$$4\frac{4}{7} - 1\frac{1}{4} = \frac{32}{7} - \frac{5}{4} = \frac{32 \times 4}{28} - \frac{5 \times 7}{28} = \frac{93}{28} = 3\frac{9}{28}$$

Key Point

Always write the 'remainder' as a fraction.

Quick Test

Work out:

1. $\frac{4}{5} \times \frac{5}{12}$

2. $\frac{7}{12} \div \frac{3}{7}$

3. $\frac{12}{15} \div \frac{5}{35}$

4. $5\frac{5}{6} + 3\frac{5}{12}$

5. $4\frac{3}{10} - 2\frac{1}{9}$

Key Words

mixed number
improper fraction

Algebra

Coordinates and Graphs 1

You must be able to:

- Understand and use coordinates
- Plot linear graphs
- Understand the components of $y = mx + c$.

Coordinates

- **Coordinates** are usually given in the form (x, y) and they are used to find certain points on a graph with an x-axis and a y-axis.

Plot the coordinates (4, 7) and (–3, 4).

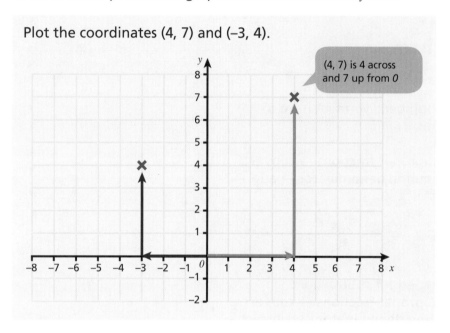

(4, 7) is 4 across and 7 up from 0

> **Key Point**
>
> The x-axis is always the one that goes across the page. The y-axis always goes up the page.
>
> When reading or plotting a coordinate, we use the first number as the position on the x-axis and the second number as the position on the y-axis.

- You use the same idea to read a coordinate from a graph or chart.

What are the coordinates of each corner of this square?

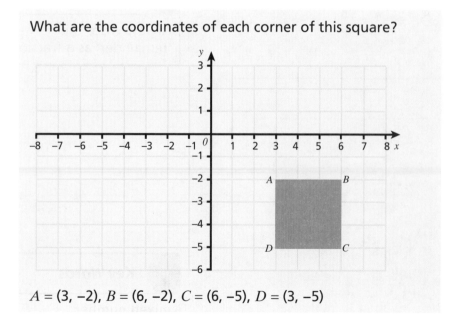

$A = (3, -2)$, $B = (6, -2)$, $C = (6, -5)$, $D = (3, -5)$

Linear Graphs

- **Linear** graphs form a straight line.
- When plotting a graph, we need to have a rule for what we are plotting, usually an equation.

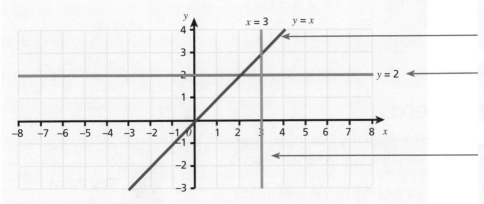

When plotting $y = x$, whatever x-coordinate you choose, the y-coordinate will be the same.

When plotting $y = 2$, you can choose any value for the x-coordinate but the y-coordinate must always equal 2

When plotting $x = 3$, you can choose any value for the y-coordinate but the x-coordinate must always equal 3

Graphs of $y = mx + c$

- m is the **gradient** of the graph.
- c is the **intercept** with the y-axis.
- To create the graph you substitute real numbers for x and y.

Plot the graph $y = 2x + 1$

If $x = 1$, you can work out y: $y = 2 \times 1 + 1 = 3$

Work out other values of y by changing the value of x.

x	−1	0	1	2	3
y	−1	1	3	5	7

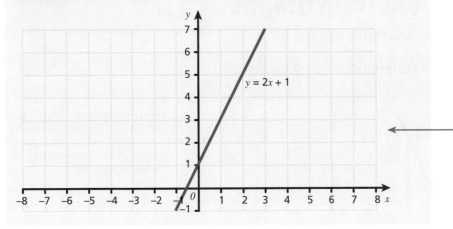

> **Key Point**
>
> A positive value of m will give a positive gradient. The graph will appear 'uphill' from left to right.
>
> A negative value of m will give a negative gradient. The graph will appear 'downhill' from left to right.

Points on the line satisfy the equation $y = 2x + 1$
Points **above** the line satisfy the inequality $y > 2x + 1$, e.g. (1, 6)
Points **below** the line satisfy the inequality $y < 2x + 1$, e.g. (3, 2)

Quick Test

1. Complete the table of values for $y = 3x - 5$

x	−2	−1	0	1	2	3
y						

> **Key Words**
>
> coordinates
> linear
> gradient
> intercept

Coordinates and Graphs 2

Quick Recall Quiz

You must be able to:

- Understand gradients and intercepts
- Solve equations from linear graphs
- Plot quadratic graphs.

Gradients and Intercepts

- You can work out the equation of a graph by looking at the gradient (how steep it is) and the y-intercept (where it crosses the y-axis).
- The gradient can be worked out by picking two points on the graph, finding the difference between the points on both the y- and x-axes and dividing them.

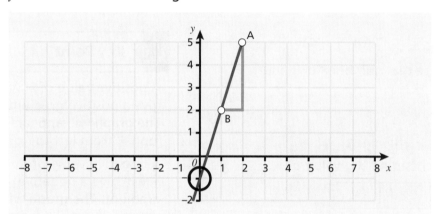

> **Key Point**
>
> gradient =
>
> $$\frac{\text{difference in } y}{\text{difference in } x}$$

When x increases by 1, y increases by 3

$3 \div 1 = 3$ so the gradient is 3

The line crosses the y-axis at −1 so the intercept is −1

The equation is $y = 3x - 1$

Solving Linear Equations from Graphs

- Graphs can be used to solve linear equations visually.

You can find the solution to the equation $2x + 3 = 7$ by using the graph $y = 2x + 3$

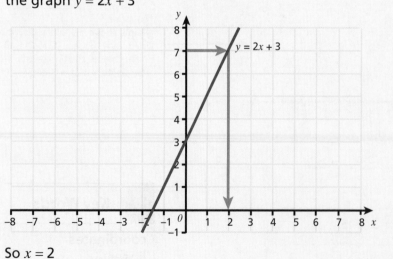

$y = 2x + 3$

First plot the graph. Then find where $y = 7$ on the y-axis. Trace your finger across until it meets the graph. Finally follow it down to read the x-axis value.

So $x = 2$

- **Simultaneous equations** are two equations that are linked. The solution to both equations can be seen at the point where their graphs cross.

> **Key Point**
>
> You can check your solution from a graph by substituting back into the equation.
>
> For example, if $x = 2$
> $y = 2x + 3 = 2(2) + 3 = 7$

The graph shows the lines $y = x + 1$ and $y = 2$

Solve the simultaneous equations $y = x + 1$ and $y = 2$

Solution is the coordinates of the point of intersection, so $x = 1$ and $y = 2$

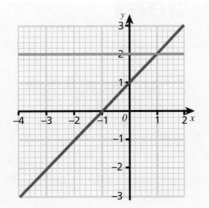

Drawing Quadratic Graphs

- **Quadratic equations** make graphs that are not linear but curved.

The equation $y = x^2$ has a power in it, so this alters the graph to one that has a curve.

x	−3	−2	−1	0	1	2	3
$y = x^2$	9	4	1	0	1	4	9

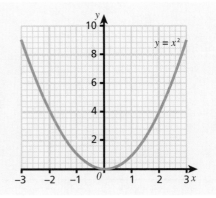

- Quadratics can take more complicated forms, but you still just substitute x for a real number to get the coordinates.

$y = x^2 + 2x + 1$

If $x = -3$

$y = (-3)^2 + 2(-3) + 1$

$= 9 + -6 + 1 = 4$

Work out other values of y by changing the value of x.

x	−3	−2	−1	0	1	2
y	4	1	0	1	4	9

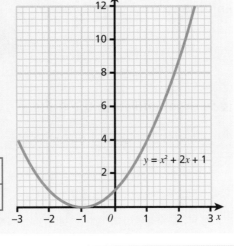

Key Point

Remember that when multiplying a negative by another negative, it becomes a positive number.

Quick Test

1. What are the gradient and y-intercepts of these equations?
   a) $y = 3x + 5$ b) $y = 6x - 7$ c) $y = -3x + 2$
2. Fill in the table for the coordinates of $y = x^2 + 3x + 4$

x	−3	−2	−1	0	1	2	3
y							

Key Words

simultaneous equations
quadratic equation

Review Questions

3D Shapes: Volume and Surface Area

1   Name the shape that has five faces, five vertices and eight edges.   [1]

2   Draw the nets of these shapes.   [3]

    a) 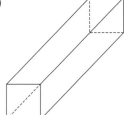   **b)** 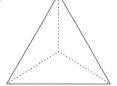   **c)**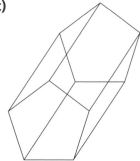

Total Marks _____ / 4

---

1   Work out the surface area and the volume of this cuboid.
 Do not forget the units.   [4]

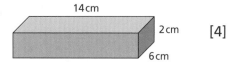

14 cm   2 cm   6 cm

2   Work out the volume of this oil drum and state the units.   [3]

2.2 m   11 m

(MR) 3   If the volume of a cube is $512 \, \text{m}^3$, what is the length of the sides in centimetres?   [2]

(MR) 4   Parveen has $1100 \, \text{cm}^2$ of paper to wrap this present.

    Does he have enough paper? Show your working.   [2]

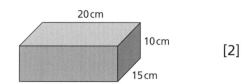

20 cm   10 cm   15 cm

Total Marks _____ / 11

Interpreting Data

1 Lalana works at a call centre. The table shows the calls she took in one day.

Call length (t minutes)	Frequency
$0 < t \leqslant 2$	25
$2 < t \leqslant 4$	40
$4 < t \leqslant 6$	18
$6 < t \leqslant 8$	10
$8 < t \leqslant 10$	4

Draw a frequency diagram for the data. [3]

Total Marks _____ / 3

(MR) **1** Plot the following data on a scatter graph. Discuss any patterns and explain what the graph shows. [4]

TV viewing figures (in 1000s)	50	45	25	65	80	75	40	30	55
TV advert spend (in £1000s)	40	30	10	45	60	70	35	15	30

2 For each survey question below, state two things that could be improved.

a) Do you eat a lot of junk food?　　　Yes ☐　　No ☐

b) How much fruit do you eat in a week?　　1 ☐　　2 ☐　　3 ☐　　4 ☐ [4]

Total Marks _____ / 8

Fractions

1. Using the grids below, prove that $\frac{3}{5}$ is smaller than $\frac{2}{3}$

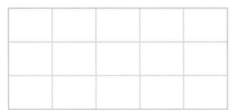 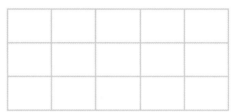

[3]

2. Work out each calculation. Simplify your answers.

   a) $\frac{4}{10} + \frac{1}{4} + \frac{2}{5} =$

   b) $\frac{2}{5} + \frac{1}{8} + \frac{1}{2} =$

   c) $\frac{3}{4} + \frac{2}{5} + \frac{3}{10} =$

   d) $\frac{3}{4} - \frac{1}{8} =$

   e) $\frac{5}{6} - \frac{1}{5} - \frac{1}{3} =$

   f) $\frac{7}{9} - \frac{1}{4} =$ [6]

3. Work out each calculation. Simplify your answers.

   a) $\frac{1}{8} \times \frac{2}{3} =$

   b) $\frac{5}{6} \times \frac{8}{9} =$

   c) $\frac{3}{10} \times \frac{1}{2} =$ [3]

4. Work out each calculation. Simplify your answers.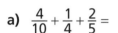

   a) $\frac{1}{8} \div \frac{2}{3} =$

   b) $\frac{1}{6} \div \frac{8}{9} =$

   c) $\frac{3}{4} \div \frac{3}{7} =$ [3]

Total Marks _____ / 15

1. Work out each calculation. Give your answers as mixed numbers.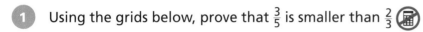

   a) $4\frac{3}{8} + 2\frac{1}{5} =$ [2]

   b) $3\frac{3}{5} + 2\frac{3}{9} =$ [2]

   c) $7\frac{1}{4} - 2\frac{8}{11} =$ [2]

   d) $2\frac{1}{5} - 1\frac{3}{7} =$ [2]

Total Marks _____ / 8

Coordinates and Graphs

1 The two points on this grid are the opposite vertices (corners) of a square.

Find the coordinates of the missing vertices.

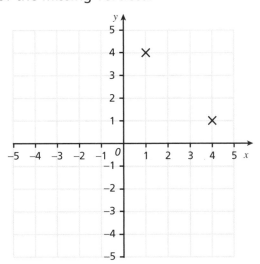

[2]

2 Draw a graph that shows the lines $x = -3$ and $y = 2$ [2]

3 Copy and complete the table of values for the equation $y = 3x - 4$

Plot your results on a graph.

[3]

x	−1	0	1	2	3	4
y						

Total Marks .... / 7

1 Copy and complete the table of values below for the quadratic equation $y = x^2 + 3x - 2$

[3]

x	−3	−2	−1	0	1	2	3
y							

Total Marks .... / 3

Angles 1

You must be able to:

- Measure and draw angles
- Use properties of triangles to solve angle problems
- Use properties of quadrilaterals to solve angle problems
- Calculate angles at a point, on a straight line and at a right angle
- Bisect an angle.

How to Measure and Draw an Angle

- Angles are measured in **degrees**.
- A **protractor** is used to measure and draw an angle.

Remember:
- angles less than 90° are called acute angles
- angles between 90° and 180° are called obtuse angles
- angles greater than 180° are called reflex angles.

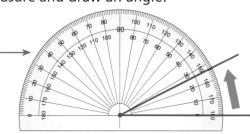

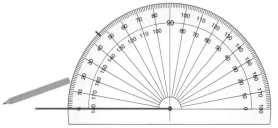

> **Key Point**
>
> There are two scales on your protractor. This is so the protractor can be used in two different directions. Always start at zero and count up.

To measure the angle of a line:
- Place the protractor with the zero line on the base line.
- The centre should be level with the point where the two lines meet.
- Counting up from zero, count the degrees of the angle you are measuring.

To draw an angle:
- Draw a base line for the angle.
- Line up your protractor, putting the centre on one end of the line.
- Count up from zero until you reach your angle, e.g. 45°
- Put a mark. Remove the protractor and draw a straight line joining the end of the base line to your point.

Angles in Triangles and Quadrilaterals

- Angles in any **triangle** add up to 180°. You can use this fact to help you work out unknown angles.
- An isosceles triangle has two equal sides and two (base) angles that are equal.
- A right-angled triangle has a 90° angle, so if you have one more angle you can work out the remaining one.

Find the size of angle x in these triangles.

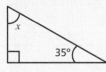

$$180° - (90° + 35°) = x = 55°$$

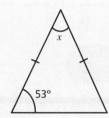

$$180° - (53° × 2) = x = 74°$$

- Angles in any **quadrilateral** add up to 360°

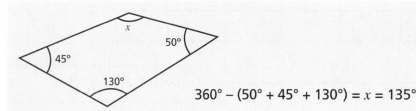

$$360° - (50° + 45° + 130°) = x = 135°$$

> **Key Point**
>
> Each triangle or quadrilateral will present properties that are useful when working out unknown angles.

- A parallelogram has two sets of equal angles. The opposite angles are equal.
- Only one angle is needed to be able to work out the others.

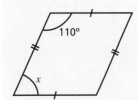

$360° - (110° \times 2) = 140°$
Now share this value equally between the remaining two angles:
$x = 140° \div 2 = 70°$

Calculating Angles

- **Angles at a point** add up to 360°
- **Angles on a straight line** add up to 180° and are called **supplementary angles**.
- **Angles in a right angle** add up to 90°
- **Vertically opposite angles** are equal.

Work out the lettered angle in each diagram.

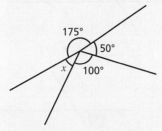

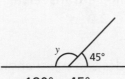

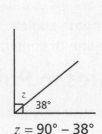

$x = 360° - 175° - 100° - 50°$
$x = 35°$

$y = 180° - 45°$
$y = 135°$

$w = 28°$

$z = 90° - 38°$
$z = 52°$

Bisecting an Angle

- **Bisect** means to cut exactly into two.

1. Open your compass.
2. Put the point of the compass on the vertex.
3. Make an arc intersecting both lines.
4. Now put the point on the first intersection and make an arc between the lines.
5. Repeat for the other intersection keeping the same radius. Make sure the arcs intersect.
6. Draw a line through the vertex and the intersection.

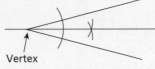

Vertex

The vertex is where the lines meet.

The vertex is where the lines meet.

Revise

Quick Test

1. Using a protractor, draw an angle of:
   a) 48°   b) 84°   c) 125°   d) 167°
2. Find the size of the marked angle in these shapes.
   a)   b)   c)

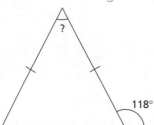
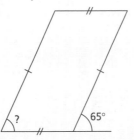

Key Words

degree
protractor
triangle
quadrilateral
supplementary angles
vertically opposite angles
bisect

Angles 2

You must be able to:

- Understand and calculate angles in parallel lines
- Use properties of a polygon to solve angle problems
- Use properties of some polygons to tessellate them.

Angles in Parallel Lines

- Parallel lines are lines that run at the same angle.
- Using parallel lines and a line that crosses them, you can apply some observations to help find missing angles.
- The angles represented by are equal. They are called **corresponding angles**.
- The angles represented by 💟 are also equal. They are called **alternate angles**.
- The line crossing the two parallel lines is called a **transversal**.

Corresponding angles

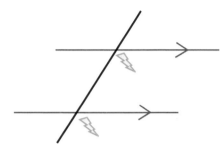

Alternate angles

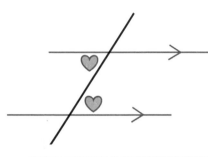

Angles in Polygons

- A **regular polygon** is a shape that has equal sides and equal angles.
- Using the fact that angles in a triangle add up to 180°, you can split any shape into triangles to help you work out the number of degrees inside that shape.

  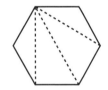

- Dividing the total number of degrees inside a regular polygon by the number of vertices will give the size of one **interior angle**.
- The sum of the **exterior angles** of any shape always equals 360°

Key Point

A polygon can be split into a number of triangles. Try this formula to speed up calculation:

(No. of sides − 2) × 180 = sum of interior angles

A regular polygon has interior angles of 140°. How many sides does it have?

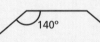

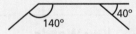

The exterior angle of each part of the polygon is
180° − 140° = 40°

360° ÷ 40° = 9, so the polygon has nine sides. ◄ A polygon with nine sides is called a nonagon.

Shape	Number of sides	Sum of the interior angles
Triangle	3	180°
Quadrilateral	4	360°
Pentagon	5	540°
Hexagon	6	720°
Heptagon	7	900°
Octagon	8	1080°
Decagon	10	1440°

Properties of Triangles and Quadrilaterals

- Here is a reminder of some commonly used shapes.

Types of triangle	**Isosceles triangle** (two equal sides, two equal angles)	Isosceles triangle	**Equilateral triangle** (three equal sides, three equal angles)	Equilateral triangle
Types of quadrilateral	**Square** (four equal sides, four right angles)	Square	**Kite** (two pairs of adjacent equal sides, one pair of equal angles)	Kite
	Rectangle (opposite sides equal, four right angles)	Rectangle	**Rhombus** (opposite sides parallel, four equal sides, opposite angles equal, diagonals bisect at right angles)	Rhombus
	Parallelogram (opposite sides equal, opposite angles equal)	Parallelogram	**Arrowhead** or **Delta** (two pairs of adjacent equal sides, one pair of equal angles)	Arrowhead
	Trapezium (one pair of parallel sides)	Trapezium		

Polygons and Tessellation

- **Tessellation** is where you repeat a shape or a number of shapes so they fit without any overlaps or gaps.

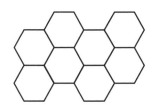

> **Quick Test**
>
> 1. Find the size of the marked angles.
>   a) 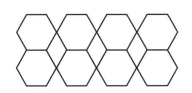 55° ?   b) ? 112°   c) ? 126°
> 2. Name a regular shape that tessellates.

Key Words

corresponding angles
alternate angles
transversal
regular polygon
interior angle
exterior angle
tessellation

Probability 1

You must be able to:

- Recognise and use words associated with probability
- Construct and use a probability scale
- Calculate the probability of an event not occurring
- Construct and use sample spaces.

Probability Words

- Certain words are used to describe the chance of an **outcome** happening. How would you describe the chance of:
  - there being 40 days in a month?
    Impossible – there are at most 31 days in a month.
  - a student attending school tomorrow?
    Likely – it cannot be said to be certain as the student might be on school holiday or ill and not attending school.
  - rolling a 2 on a dice?
    Unlikely – there are six possible outcomes and the number 2 is only one of these.
  - taking a green sweet from a bag that only contains green sweets?
    Certain – in this case there is no other outcome possible.
  - flipping a fair coin and it landing on heads?
    An **even chance** – the outcome could be a head or a tail, two equally likely options.

> ### Key Point
>
> Try to consider the event with all the possible outcomes.
>
> Once all the possible outcomes have been considered, the word can be selected.

Probability Scale

- The **probability** of a particular outcome can be described using a **probability scale** from 0 to 1
- Probabilities are written as fractions, decimals or percentages.
- The probability of an outcome happening is written as

$$P(\text{outcome}) = \frac{\text{number of ways the outcome can happen}}{\text{total number of all possible outcomes}}$$

(unlikely)   (likely)

0   0.5   1
(impossible)   (even chance)   (certain)

Events and Outcomes

- An **event** is a set of outcomes, e.g. rolling a dice.
- An event is **fair** when the outcomes are **equally likely**.
- An event is **biased** when the outcomes are **not** equally likely.
- **Random** means each possible outcome is equally likely.

When you throw a fair coin:

$P(\text{Head})$ or $P(H) = \frac{1}{2}$ and $P(\text{Tail})$ or $P(T) = \frac{1}{2}$

⟵ Use this notation to save time.

A bag contains 5 red and 4 blue counters. One counter is taken from the bag at random. What is the probability the counter is:

a) red?   Five of the counters are red, so P(red) = $\frac{5}{9}$

b) blue?   Four of the counters are blue, so P(blue) = $\frac{4}{9}$

c) green?   There are no green counters, so P(green) = 0

- The probabilities of all possible outcomes sum to 1. In the example above, P(red or blue) = $\frac{9}{9}$ = 1

Probability of an Outcome Not Happening

- The probability of an outcome **not** happening is:
  1 – the probability of the outcome happening

A fair six-sided dice is rolled.

P(rolling a 2) = $\frac{1}{6}$

P(**not** rolling a 2) = $1-\frac{1}{6}=\frac{5}{6}$

P(rolling a 4 or a 5) = $\frac{2}{6}$

P(**not** rolling a 4 or a 5) = $1-\frac{2}{6}=\frac{4}{6}$

There are five outcomes that are not a 2; these are 1, 3, 4, 5, 6

The probability of the weather being cloudy = 0.4

So the probability of it **not** being cloudy is 1 – 0.4 = 0.6

> **Key Point**
>
> The sum of the probabilities of all possible outcomes is 1

Sample Spaces

- A **sample space** shows all the possible outcomes of the event.
- When two or more events take place, they are called **combined events**.

The sample space for throwing a coin and rolling a dice is:

H1	H2	H3	H4	H5	H6
T1	T2	T3	T4	T5	T6

This shows all the possible outcomes and helps us to calculate probabilities.

P(H1) = $\frac{1}{12}$

> **Key Words**
>
> outcome
> impossible
> likely
> unlikely
> certain
> even chance
> probability
> probability scale
> event
> fair
> equally likely
> biased
> random
> sample space
> combined events

> **Quick Test**
>
> 1. In London, how can the chance of rain in October be described?
> 2. Show the outcome in question 1 on a probability scale.
> 3. A bag contains 5 red sweets and 10 blue sweets.
>   a) What is the probability of picking a red sweet?
>   b) What is the probability of not picking a red sweet?
> 4. If it rains 0.75 of the time in a rainforest and is cloudy 0.1 of the time, what is the probability it isn't raining or cloudy?

Probability 2

You must be able to:

* Understand what mutually exclusive outcomes are
* Calculate a probability with and without a table
* Work with experimental probability
* Understand Venn diagrams and set notation.

Mutually Exclusive Outcomes

* **Mutually exclusive** outcomes are outcomes that cannot happen at the same time.
* For example, the arrow on the spinner **cannot** land on yellow and blue at the same time. Because the outcomes are mutually exclusive, P(yellow and blue) = 0
* You can calculate the probability of spinning yellow or blue as
P(yellow or blue) = P(yellow) + P(blue) = $\frac{1}{4} + \frac{1}{4} = \frac{1}{2}$

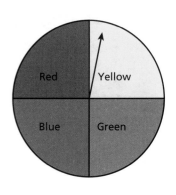

Probability Calculations

* Probabilities can be used to predict possible outcomes.
* If the outcome of one event does **not** affect the outcome of another event, they are **independent**.
* If the outcome of one event does affect the outcome of another event, they are **dependent**.
* **Conditional probability** is when the probability is affected by a previous outcome.

> **Key Point**
>
> When outcomes are mutually exclusive, using the word **or** implies we can add the probabilities of the outcomes together.

What is the probability of picking a green or black ball, at random, from the bag?

P(green or black) = $\frac{3}{15} + \frac{4}{15} = \frac{7}{15}$

If a red ball is removed from the bag, what is the probability the next ball is green?

There are only 14 balls left so P(green) = $\frac{3}{14}$

This is conditional probability.

This bag contains 15 balls: 3 are yellow, the rest are either red or blue. Use the table of probabilities to work out the number of each colour ball.

Yellow	Red	Blue
0.2	0.6	?

The probability that any ball is chosen is 1 so we can calculate the probability of drawing a blue ball: 1 − 0.6 − 0.2 = 0.2

There are the same number of blue as yellow as both have probability 0.2

One way of working out the actual number of each colour is to multiply the probability by the total number of balls:
15 × 0.6 = 9 (red)   0.2 × 15 = 3 (blue)

There are three times as many red as yellow as 0.6 = 3 × 0.2

Experimental Probability

- The probability of a 6 when rolling a fair six-sided dice is $\frac{1}{6}$
  If you roll the dice six times, would you definitely get a 6? You may not get a 6 in six rolls. However, the more times you roll the dice, the more likely you are to get closer to a probability of $\frac{1}{6}$
  This is **experimental probability**.

Hayley sat outside her school and counted 25 cars that went past. She noted the colour of each car in this table.

Yellow	1
Red	6
Blue	4
Black	9
White	5

a) What is the probability of the next car going past being white?

$\frac{5}{25} = \frac{1}{5} = 0.2$

b) How many black cars would you expect if 50 cars go past?

$\frac{9}{25} = 0.36$ \qquad $0.36 \times 50 = 18$

Venn Diagrams and Set Notation

- Venn diagrams can be used to organise sets and find probabilities.

Set A is the children who like green beans. Set B is the children who like carrots.

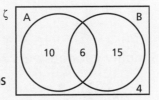

(A) = 10 + 6 = 16 children like green beans

(B) = 6 + 15 = 21 children like carrots

(A ∪ B) = 10 + 6 + 15 = 31 children like at least one. $\longleftarrow$ This is the union of A and B.

(A ∩ B) = 6 children like both green beans and carrots. $\longleftarrow$ This is the intersection of A and B.

(A ∪ B)' = 4 children like neither green beans nor carrots. $\longleftarrow$ The symbol ' means not in A or B.

The probability that a child does **not** like green beans or carrots = $\frac{4}{35}$

1. A bag has 20 balls of four different colours.
   a) Complete the table of probabilities.
   b) What is the probability of **not** getting a yellow ball?
   c) A ball is chosen at random and replaced 60 times. What is the expected number of times a green ball is chosen?

Yellow	?
Red	0.3
Green	0.2
Blue	0.1

Fractions

1 What fraction of each shape is shaded in? Give two other equivalent fractions for each. [3]

a)   b)   c)

Total Marks .... / 3

1 Work out each calculation. Simplify your answers where possible.

a) $\frac{2}{5} + \frac{1}{10} =$ b) $\frac{7}{12} + \frac{1}{4} =$ c) $\frac{1}{6} + \frac{1}{5} =$

d) $\frac{2}{7} + \frac{3}{10} =$ e) $\frac{8}{9} - \frac{1}{3} =$ f) $\frac{7}{11} - \frac{1}{2} =$

g) $\frac{9}{10} - \frac{2}{3} =$ [7]

2 Work out each calculation. Simplify your answers where possible.

a) $\frac{4}{9} \times \frac{1}{5} =$ b) $\frac{3}{7} \times \frac{3}{10} =$ c) $\frac{5}{12} \times \frac{2}{3} =$

d) $\frac{2}{9} \div \frac{1}{4} =$ e) $\frac{4}{5} \div \frac{6}{11} =$ [5]

FS
MR 3 Sally won some money in the lottery. She gave $\frac{2}{5}$ to her husband and $\frac{1}{4}$ to her daughter.

What fraction did she keep? [3]

MR 4 Keith was baking a cake. His recipe was $\frac{4}{9}$ flour, $\frac{1}{3}$ sugar and butter, and the rest was an equal split of chocolate and eggs.

What fraction does the chocolate part represent? [3]

PS 5 Change the following mixed numbers to improper fractions.

a) $8\frac{5}{9}$ b) $3\frac{2}{7}$ c) $1\frac{3}{11}$ [3]

Total Marks .... / 21

Coordinates and Graphs

1 Copy the grid shown and plot the following points:

a) (1, 2) [1]

b) (−4, 5) [1]

c) (3, −2) [1]

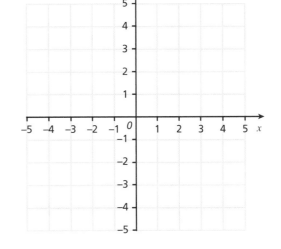

2 On the same grid plot the following lines:

a) $x = 4$ b) $y = −1$ [2]

Total Marks _____ / 5

---

1 Copy and complete the arrows and the table of values for the equation $y = 2x − 4$. Then plot your results on a grid like the one shown.

x	−1	0	1	2	3
y					

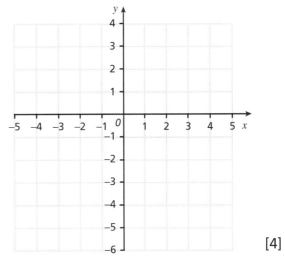

[4]

2 Copy and complete the table of values for the equation below.

$y = x^2 + 5x + 1$

x	−3	−2	−1	0	1	2	3
y							

[3]

Total Marks _____ / 7

Angles

1   Find the size of the marked angles in the following diagrams.

a)

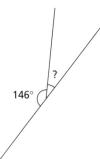

b)

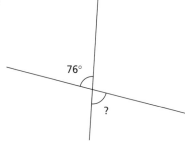

c)

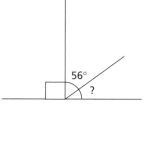

[3]

2   Find the size of the marked angle in this isosceles triangle.

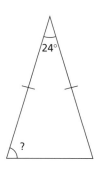

[2]

3   What is the name of a polygon with seven sides?   [1]

Total Marks .... / 6

1   Find the size of angles x and y in each diagram.

a)

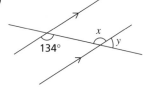

b)

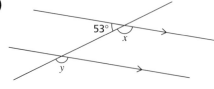

[4]

2   If each interior angle of a regular polygon is 150°, how many sides does it have?   [2]

Total Marks .... / 6

Probability

1 If today is Sunday, what is the chance that tomorrow is Tuesday? [1]

2 An eight-sided fair spinner is numbered 0, 0, 0, 1, 1, 1, 1, 2

   a) What is the probability of getting a 0 on one spin? [1]

   b) What is the probability of **not** getting a 2? [2]

Total Marks .... / 4

1 Ranjeet drops a paper cup a number of times. He finds the probability of it landing upside-down is 0.65

What is the probability of it **not** landing upside-down? [2]

2 A dice has been rolled 50 times and the score recorded in the frequency table below.

   a) Copy and complete the table. Give the probabilities as fractions.

Number	Frequency	Estimated probability
1	5	
2	8	
3	7	
4	7	
5	8	
6	15	
Total 50	**1**	

[3]

   b) Use the results in your table to work out the estimated probability (as fractions) of getting each of the following.

     i) The number 6   **ii)** An odd number   **iii)** A number greater than 4 [3]

Total Marks .... / 8

Fractions, Decimals and Percentages 1

Quick Recall Quiz

You must be able to:

- Convert between a fraction, decimal and percentage
- Order fractions, decimals and percentages
- Calculate a fraction of a quantity and a percentage of a quantity
- Compare quantities using percentages.

Fractions, Decimals and Percentages

- The table below shows how **fractions**, **decimals** and **percentages** are used to represent the same amount:

Picture	Fraction	Decimal	Percentage
$\frac{1}{4}$	$\frac{1}{4}$	0.25	25%
$\frac{1}{2}$	$\frac{1}{2}$	0.5	50%
$\frac{3}{4}$	$\frac{3}{4}$	0.75	75%

> **Key Point**
>
> **Percent** means 'out of 100'

Carrying out Conversions

- For conversions you do not know automatically, use the rules below.

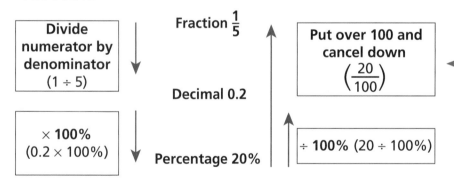

Ordering Fractions, Decimals and Percentages

- To order fractions, decimals and percentages, change them all to the same format.

Put 0.25, $\frac{2}{5}$ and 20% in order from smallest to largest.

Changing them all to decimals gives $\frac{2}{5} = 0.4$ and 20% = 0.2

0.2 < 0.25 and 0.25 < 0.4, so correct order is 20%, 0.25, $\frac{2}{5}$

Fractions of a Quantity

- To find a fraction of a **quantity** without a calculator, divide by the **denominator** and multiply by the **numerator**.

  Find $\frac{2}{3}$ of £120

  $120 \div 3 \times 2 = £80$

- To find a fraction of a quantity with a calculator, use the fraction button . Calculators can differ, so find out how yours works with fractions.

  Find $\frac{2}{5}$ of £200

  $2 \; \blacksquare \; 5 \times 200 = £80$

Percentages of a Quantity

- To find a percentage of a quantity without a calculator:

  Find 20% of £60

  10% of £60 $= 60 \div 10 = £6$
  $20\% = £6 \times 2 = £12$

  > 20% is two lots of 10%

- To find a percentage of a quantity with a calculator:

  Find 20% of £60

  $= 20\% \times £60$
  $= 20 \div 100 \times 60 = £12$

  Or $20\% = 0.2$
  so 20% of £60 is
  $0.2 \times £60 = £12$

  > Change the percentage to a decimal.

  Find 120% of £60

  $= 120\% \times £60$
  $= 120 \div 100 \times 60 = £72$

  Or $120\% = 1.2$
  so 120% of £60 is
  $1.2 \times £60 = £72$

  > Remember: as 120% > 100%, your answer will be bigger than £60

- To compare two quantities using percentages:

  Which is bigger, 30% of £600 or 25% of £700?

  10% of £600 $= 600 \div 10 = £60$ 50% of £700 $= 700 \div 2 = £350$
  30% of £600 $= 60 \times 3 = £180$ 25% of £700 $= 350 \div 2 = £175$

  So 30% of £600 is bigger, by £5

> ### Key Point
>
> Useful percentages to know:
>
> $50\% \rightarrow \div 2$
>
> $10\% \rightarrow \div 10$
>
> $1\% \rightarrow \div 100$
>
> 'of' means '$\times$'

Quick Test

1. Change $\frac{7}{20}$ to a decimal and a percentage.
2. Change 36% to a fraction in its simplest form.
3. Work out $\frac{2}{5}$ of £70
4. Find 35% of $140

Key Words

fraction
decimal
percentage
quantity

Fractions, Decimals and Percentages 2

You must be able to:

- Increase or decrease a quantity by a percentage
- Work out a percentage change
- Find one quantity as a percentage of another
- Work out simple interest and tax.

Increasing and Decreasing Quantities by a Percentage

- To **increase** or **decrease** a quantity by a percentage, add on or subtract the percentage you have found.
- You can also use a **multiplier**.

> A calculator is priced at £12 but there is a discount of 25%
> Work out the reduced price of the calculator.
>
> 25% of £12 is one-quarter of £12 = £3
> Reduced price = £12 − £3 = £9
> **Using a multiplier:**
> A reduction of 25% means you are left with 75%, and 75% = 0.75, so the multiplier is 0.75
> 0.75 × £12 = £9

£3 is the discount so **'take it away'** to get the final answer.

> A laptop computer costs £350 plus tax at 20%
> Work out the actual cost of the laptop.
> 20% of £350 is 20% × £350 = 20 ÷ 100 × 350
> $\qquad\qquad\qquad\qquad\qquad\qquad$ = £70
>
> Actual cost = £350 + £70 = £420
> **Using a multiplier:**
> An increase of 20% means you pay 120%, and 120% = 1.2
> 1.2 × £350 = £420

£70 is the tax so **'add it on'** to get the final answer.

So the multiplier is 1.2

Percentage Change

- You can work out the percentage change using

$$\text{Percentage change} = \frac{\text{change}}{\text{original}} \times 100\%$$

> The number of students in a class increases from 25 to 30
> Work out the percentage increase.
>
> Percentage increase is $\frac{5}{25} \times 100\% = 20\%$

> In a sale, a coat is reduced from £50 to £30
> Work out the percentage reduction.
>
> Percentage reduction is $\frac{20}{50} \times 100\% = 40\%$

One Quantity as a Percentage of Another

Jane gets 18 out of 20 in a test. What percentage is this?

With a calculator:

$$\frac{18}{20} \times 100\%$$
$$= 18 \div 20 \times 100\%$$
$$= 90\%$$

Without a calculator:

$$\frac{18}{20} = \frac{90}{100} = 90\%$$

$\times 5$ ↓ ... $\times 5$ ↑

← Make the fraction 'out of' 100

Simple Interest and Tax

* Find the **interest** for one year then multiply by the number of years.

> Peter puts £200 into a savings account. He gets 5% simple interest per year. How much does he have in his account after two years?
>
> $$10\% \text{ of } £200 = 200 \div 10 = £20$$
> $$5\% = 20 \div 2 = £10$$
>
> After two years Peter receives £10 × 2 = £20
>
> Peter has £200 + £20 = £220 in his account.

← This is the interest for one year.

> Samira borrows £2000 from her bank. She has to pay back the loan with additional interest at 8% per year. After three years, how much will Samira have to pay back?
>
> $$8\% \text{ of } £2000 = 8\% \times £2000$$
> $$= 8 \div 100 \times 2000 = £160$$
>
> After three years the interest will be £160 × 3 = £480
>
> So Samira will have to pay back £2000 + £480 = £2480

← This is the interest for one year.

> **Key Point**
>
> Simple interest is **not** added on at the end of each year.

* You have to pay tax on money you earn, called **income tax**, and also on some things you buy, called **value added tax (VAT)**.

> A man earns £30 000 per year. The first £12 500 is tax free. He pays 20% income tax on the rest. How much does he pay?
>
> He pays tax on £30 000 – £12 500 = £17 500
> 20% of 17 500 = 20 ÷ 100 × 17 500 = £3500

> A plumber charges £60 per hour plus VAT. VAT is 20% How much does he charge including VAT?
>
> VAT is 20% so 100% + 20% = 120% = 1.2
> 1.2 × £60 = £72

← So the multiplier is 1.2

> **Key Words**
>
> increase
> decrease
> multiplier
> interest
> income tax
> value added tax (VAT)

> **Quick Test**
>
> 1. A television costing £450 is reduced by 10% What is its sale price?
> 2. A house costing £80 000 increases in value by 15% What is the value of the house now?
> 3. Anne gets $\frac{21}{25}$ in a test. What percentage is this?

Equations 1

You must be able to:

- Find an unknown number
- Solve a simple equation
- Solve an equation with unknowns on both sides
- Apply the inverse of an operation.

Finding Unknown Numbers

- The **unknown** number is usually given as a letter or symbol.
- Applying the **inverse** operations helps you to find the unknown number.

$n + 4 = 13$ or $\blacksquare + 4 = 13$

← Both ways mean **'something + 4 = 13'**

Take 4 away from 13

n or $\blacksquare = 13 - 4$

← -4 is the 'inverse' or 'opposite' of $+4$

n or $\blacksquare = 9$ (check: $9 + 4 = 13$)

$x - 4 = 13$ or $\bullet - 4 = 13$

← Both ways mean **'something − 4 = 13'**

Add 4 to 13

x or $\bullet = 13 + 4$

← $+4$ is the 'inverse' or 'opposite' of -4

x or $\bullet = 17$ (check: $17 - 4 = 13$)

- The same idea applies to multiplying and dividing.

$3n = 12$

← $3 \times$ something $= 12$

$n = 12 \div 3$

← $\div 3$ is the inverse of $\times 3$

$n = 4$ (check: $3 \times 4 = 12$)

$\frac{n}{3} = 4$

← Something $\div 3 = 4$

$n = 4 \times 3$

← $\times 3$ is the inverse of $\div 3$

$n = 12$ (check: $12 \div 3 = 4$)

- Now you can apply inverse operations to **solve** more difficult equations.

Key Point

Operation	Inverse
+	−
−	+
×	÷
÷	×

Solving Equations

- Remember to think of the letter as 'something'.

Solve the equation $2y + 3 = 15$

This simply means 'something' $+ 3 = 15$ ← 'Something' must be 12

So $2y = 12$ means $2 \times$ 'something' $= 12$ ← 'Something' must be 6

So $y = 6$

Now look with the **inverses**:

$$2y + 3 = 15$$
$$(-3) \quad 2y = 12$$
$$(\div 2) \quad y = 6$$

Key Point

Remember to do the same to both sides of the equation.

Equations with Unknowns on Both Sides

- An equation may have an unknown number on both sides of the equals sign.

Solve the equation $5x - 2 = 3x + 5$

$$5x - 2 = 3x + 5$$ ← Do the same to both sides.
$$(-3x) \quad 2x - 2 = 5$$ ← Subtract $3x$ from both sides so that the x term is on the left-hand side only.
$$(+2) \quad 2x = 7$$
$$(\div 2) \quad x = 3.5$$ ← Now do the inverses.

Solve the equation $3x + 5 = 5x - 4$

$$3x + 5 = 5x - 4$$
$$(-3x) \quad 5 = 2x - 4$$
$$(+4) \quad 9 = 2x$$
$$(\div 2) \quad 4.5 = x$$ ← This is the same as $x = 4.5$

Quick Test

1. What is the inverse of $\times 6$?
2. If $- 5 = 7$, what is the value of ?
3. If $6n = 30$, what is the value of n?
4. Solve the equation $3y - 2 = 13$
5. Solve the equation $3x + 7 = 2x - 2$

Key Words

unknown
solve

Equations 2

You must be able to:

- Solve equations with fractions or negative numbers
- Set up and solve an equation
- Apply the inverse of an operation.

Solving More Complex Equations

- An equation may include a **negative** of the unknown number.
- The unknown number may be part of a fraction or inside **brackets**.

Solve the equation $3x + 1 = 11 - 2x$

$$3x + 1 = 11 - 2x$$

$(+ 2x)$ $5x + 1 = 11$

$(- 1)$ $5x = 10$

$(\div 5)$ $x = 2$

Adding $2x$ to both sides so the x term is on the left-hand side only.

Solve the equation $\dfrac{3x + 1}{2} = 8$

$$\dfrac{3x + 1}{2} = 8$$

$(\times 2)$ $3x + 1 = 16$

$(- 1)$ $3x = 15$

$(\div 3)$ $x = 5$

Here the $\times 2$ cancels out the $\div 2$ on the left-hand side of the equation.

Solve the equation $3(2x - 1) = 4(x + 2)$

Multiply out the brackets first then solve in the usual way:

$$3(2x - 1) = 4(x + 2)$$

$$6x - 3 = 4x + 8$$

$(- 4x)$ $2x - 3 = 8$

$(+ 3)$ $2x = 11$

$(\div 2)$ $x = 5.5$

Key Point

Remember to multiply everything inside the bracket by the number outside the bracket.

Setting Up and Solving Equations

- You will have to follow a set of instructions in a given order.
- Usually you only have to ×, ÷, +, − or square.
- If you have to multiply 'everything' then remember to use brackets.

A number is **doubled**, then 5 is added to the total and the result is 11

What was the original number?

The words	The algebra
A number	n
doubled	$2n$
add 5	$2n + 5$
the result is 11	$2n + 5 = 11$

You can now solve this in the usual way to find the original number equals 3

Three boys were paid £10 per hour plus a tip of £6 to wash some cars. They shared the money and each got £12

How many hours did they wash cars for?

£10 × number of hours   £6 tip

Set up an equation: $\dfrac{10x + 6}{3} = 12$ each got

shared by 3 boys

Now solve the equation:

$$(\times 3) \quad 10x + 6 = 36$$
$$(-6) \quad\quad 10x = 30$$
$$(\div 10) \quad\quad x = 3$$

The boys worked for 3 hours.

Key Point

Remember to do the same to both sides of the equation.

Quick Test

1. Solve the equation $3x - 4 = 11$
2. Solve the equation $2x + 3 = 12 - x$
3. Solve the equation $2(x + 2) = 2(3x - 2)$
4. A number is multiplied by 3 and then 8 is subtracted. The result is 25. What is the number?
5. A chicken is roasted for 60 minutes for every kilogram, plus an extra 20 minutes. If the chicken took 140 minutes to cook, how heavy was it?

Key Words

negative
brackets
double

Review Questions

Angles

1 Find the size of the marked angles.

a)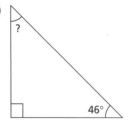

b)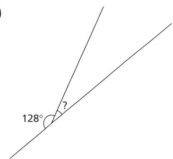

c)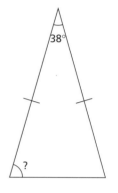

[3]

Total Marks _____ / 3

1 Find the size of the lettered angles in the parallel line diagrams below.

a)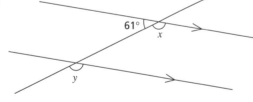

b)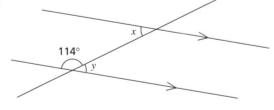

[4]

2 What do the total interior angles add up to in a nonagon? [1]

3 If each interior angle of a regular polygon equals 160°, how many sides does it have?

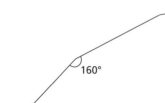

160°

[2]

Total Marks _____ / 7

Probability

1 Copy the probability scale and complete the boxes with suitable chance words.

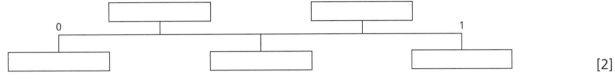

[2]

2 Dev has a drawer of 10 t-shirts: 3 blue, 4 yellow, 2 red and 1 pink. What is the probability that:

a) Dev pulls out a blue t-shirt? [1]

b) Dev pulls out either a yellow t-shirt or a red t-shirt? [1]

c) Dev does not pull out a pink t-shirt? [2]

Total Marks _____ / 6

1 Leanne runs an ice-cream van. At random, she chooses which kind of sprinkles to put on the ice-creams. The table below shows the sprinkles Leanne chose on Sunday.

Sprinkles	Frequency	Probability
Chocolate	19	
Hundreds and thousands	14	
Strawberry	7	
Nuts	10	

a) Copy the table above and complete the experimental probabilities. [2]

b) What was the probability of getting either nuts or chocolate sprinkles? [2]

2 The probability of winning a raffle prize is 0.47

What is the probability of not winning a raffle prize? [1]

3 a) Copy and complete the table below.

Sales destination	Probability of going to destination
London	0.26
Cardiff	0.15
Chester	0.2
Manchester	

[1]

b) Which is the least likely destination to travel to for sales? [1]

Total Marks _____ / 7

Practice Questions

Fractions, Decimals and Percentages

1 Copy and complete the following table of equivalent fractions, decimals and percentages.

Fraction	Decimal	Percentage
$\frac{7}{10}$		
		55
	0.32	
$\frac{3}{100}$		

[4]

2 Work out:

a) 50% of £32 [1]

b) 10% of 80 cm [1]

c) 15% of 160 m [2]

d) 25% of £104 [1]

3 Use the fraction button on your calculator to work out:

a) $\frac{1}{5}$ of £85 [1]

b) $\frac{2}{3}$ of £120 [1]

c) $\frac{5}{7}$ of 21 m [1]

Total Marks _____ / 12

1 Work out the following, showing your working.

a) $\frac{2}{3}$ of £15 [3]

b) $\frac{3}{7}$ of £210 [3]

c) $\frac{4}{5}$ of £6000 [3]

(PS) **2** A jacket costing £75 is reduced by 20% in a sale. What is the sale price of the jacket? [3]

(PS) **3** In a Maths test, Karim scored 16 out of 20 and John scored 15 out of 20

What percentage did Karim and John each get? [4]

Total Marks _____ / 16

Equations

(PS) **1** Use inverses to work out the unknown number.

    a) $\triangle - 7 = 3$ [1]

    b) $18 + \triangle = 23$ [1]

    c) $\frac{n}{4} = 3$ [1]

    d) $5y = 35$ [1]

(PS) **2** Solve these equations.

    a) $3n + 1 = 13$ [2]

    b) $2x - 5 = 3$ [2]

    c) $5y + 1 = 11$ [2]

Total Marks _____ / 10

(PS) **1** Solve these equations.

    a) $3x + 1 = x + 7$ [2]

    b) $2(2x - 3) = x - 3$ [2]

    c) $6(x + 1) = 2(x + 13)$ [2]

    d) $\frac{3x + 5}{4} = 5$ [2]

(PS) **2** A number is multiplied by 3, then 2 is added to the total. The result is 11

    Set up and solve an equation to find the original number.   [2]

(PS) **3** Mo adds together his age and the age of his sister. He gets a total of 28

    If Mo is 16, how old is his sister? Set up and solve an equation to find your answer.   [2]

Total Marks _____ / 12

Symmetry and Enlargement 1

Quick Recall Quiz

You must be able to:

- Reflect a shape
- Translate a shape
- Find the order of rotational symmetry
- Rotate a shape
- Enlarge a shape.

Reflection and Reflectional Symmetry

- Reflect each point one at a time.
- Use a line that is **perpendicular** to the mirror line.
- Make sure the **reflection** is the same distance from the mirror line as the original shape.
- A shape has reflectional symmetry if you can draw a mirror line through it.

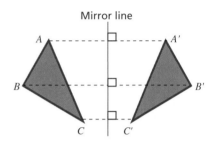

Mirror line

Translation

- Translation moves a shape left/right (x) and/or up/down (y).
- The translation is described using a column vector $\begin{pmatrix} x \\ y \end{pmatrix}$

Rotational Symmetry

- A shape has rotational symmetry if it looks exactly like the original shape when it is **rotated**.
- The **order of rotational symmetry** is the number of ways the shape looks the same.
- To rotate a shape you need to know:
  - The **centre of rotation**
  - The direction of rotation
  - The number of degrees to rotate it.

Order 1　　Order 2　　Order 3　　Order 4

> **Key Point**
>
> Perpendicular means at right angles (90°) to.
>
> Check the position of your reflection by placing a mirror along the mirror line.

a) Rotate shape A 180° about (1, 0)

b) Translate shape A by vector $\begin{pmatrix} -4 \\ 1 \end{pmatrix}$

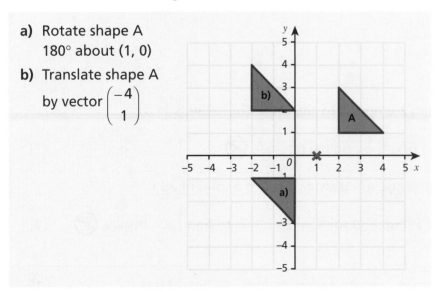

Enlargement

- To draw an **enlargement** you need to know two things:
  - How much bigger/smaller to make the shape. This is called the **scale factor**.
  - Where you will enlarge the shape from. This is called the **centre of enlargement**.
- Remember that the original and enlarged shapes are **similar** (the same shape but a different size).

Enlarge shape A by a scale factor of 2 from the point (3, 4)

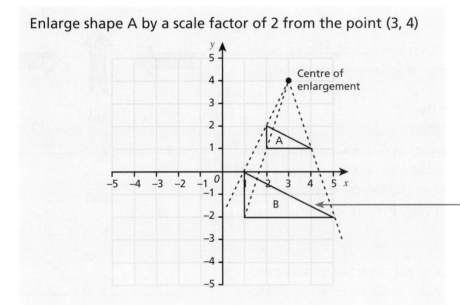

Enlarge every side of the shape.

- Use **rays** to check the position of your enlargement. They will touch corresponding corners of the shape.

Quick Test

1. What is the order of rotational symmetry of this shape?

2. a) Reflect this shape across the mirror line.
   b) Rotate the original shape 180° about the corner A.
   c) Rotate the original shape 90° clockwise about the corner B.

3. Enlarge this shape by a scale factor of 3

   1 cm
   2 cm

Key Words

reflection
rotation
centre of rotation
enlargement
scale factor
centre of enlargement
similar
ray

Symmetry and Enlargement 2

Quick Recall Quiz
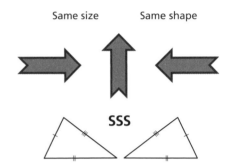

You must be able to:

- Recognise congruent shapes
- Interpret a scale drawing
- Work out side lengths in similar shapes
- Convert between units of measure
- Carry out ruler and compass constructions.

Congruence

- Congruence simply means shapes that are exactly the same.
- These arrow shapes are **congruent** – they have the same size and the same shape.
- Triangles are congruent to each other if:
  - three pairs of sides are equal (SSS)
  - two pairs of sides and the angle between them are equal (SAS)
  - two pairs of angles and the side between them are equal (ASA)
  - both triangles have a right angle, the hypotenuses are equal and one pair of corresponding sides is equal (RHS).

Scale Drawings

- A scale drawing is one that shows a real object with accurate dimensions, except they have all been reduced or enlarged by a certain amount (called the **scale**).
- Similar shapes are enlarged by the same scale factor, but the angles stay the same. All sides must be multiplied by the same value.
- A scale of 1 : 10 means in the real world the object would be 10 times bigger than in the drawing.
- We use scale drawings to represent real objects.

Shape and Ratio

- You can use **ratios** to work out the 'real' size of an object. The scale is given as a ratio with the smaller **unit** first.

Same size   Same shape

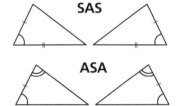

SSS

SAS

ASA

RHS

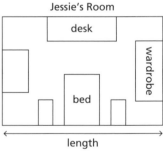
Jessie's Room

desk

wardrobe

bed

length

Scale: 1 inch = 3 feet

Here you would have to measure each side in inches then **multiply by 3** to get the real length in feet.

For every 1 cm you measure in the picture, multiply by 160 to get the real size. Then convert to metres.

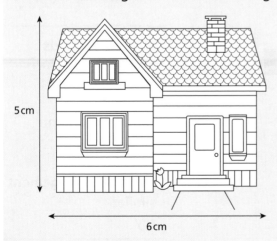
Estimate the height of this house using the scale of 1 : 160

5 cm

6 cm

Key Point

Scales are given as a ratio, usually 1 : n where n is what you multiply by.

The height of the house is 5 cm so in real life it is:
$5 \times 160 = 800$ cm $= 8$ m
The width of the house is 6 cm so in real life it is:
$6 \times 160 = 960$ cm $= 9$ m 60 cm

Remember: 100 cm = 1 m

Constructions

- You need to learn these constructions.

- The **perpendicular bisector** of line AB. 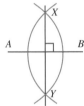	- A **perpendicular from a given point** to the line AB. 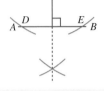
1. Open your compass to more than half AB. 2. Put the compass point on A. 3. Draw an arc extending past the centre both above and below the line. 4. Put the compass point on B and repeat Step 3 5. Draw a line through the two points where the arcs intersect. XY is the perpendicular bisector.	1. Put the compass point on C. 2. Draw an arc on both sides of the line. 3. Keeping the radius the same, put the compass point on E. 4. Draw an arc. 5. Put the compass point on D. 6. Draw an arc. 7. Join C to the point where the arcs cross.

- A **locus** is a path of a point that moves, according to a rule, so the perpendicular bisector gives the locus of all points that are the same distance from A as B.

> **Key Point**
>
> The shortest distance from a point to a line is the perpendicular distance.

> **Quick Test**
>
> 1. Draw three congruent shapes.
> 2. Which shape is congruent to shape A?
>
>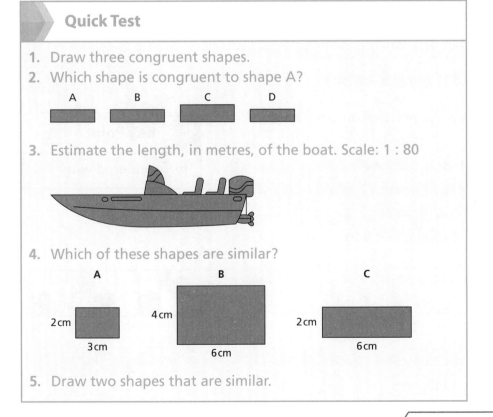
>
> 3. Estimate the length, in metres, of the boat. Scale: 1 : 80
>
>
>
> 4. Which of these shapes are similar?
>
> 5. Draw two shapes that are similar.

> **Key Words**
>
> congruent
> scale
> ratio
> units
> perpendicular bisector
> locus

Ratio and Proportion 1

You must be able to:

- Change confidently between related standard units
- Understand what ratio means
- Simplify a ratio
- Link ratios to fractions and fractions to ratios.

Changing Units

- You need to know some conversions for standard units.

Time	Length	Area	Volume	Capacity	Mass
1 minute = 60 seconds	1 cm = 10 mm	1 cm² = 100 mm²	1 cm³ = 1000 mm³	1 litre = 1000 ml	1 kg = 1000 g
1 hour = 60 minutes	1 m = 100 cm	1 m² = 10 000 cm²	1 m³ = 1 000 000 cm³	1 litre = 100 cl	1 tonne = 1000 kg
	1 km = 1000 m	1 km² = 1 000 000 m²			

Introduction to Ratios

- Ratio is a way of showing the relationship between two numbers.

> Here is a ratio table with a missing number. Work out the missing number.
>
> | 3 | 12 |
> |---|---|
> | 18 | |
>
> The horizontal multiplier is × 4   3 × 4 = 12
> The vertical multiplier is × 6   3 × 6 = 18
>
> So the missing number is 18 × 4 or 12 × 6 = 72

- Ratios can be used to compare costs, weights and sizes.

> On the deck of a boat there are 2 women and 1 man. There are also 5 cars and 2 bicycles.
>
> The ratio of men to women is 1 to 2, written 1 : 2
> The ratio of women to men is 2 to 1, written 2 : 1
> The ratio of cars to bicycles is 5 to 2, written 5 : 2
> The ratio of bicycles to cars is 2 to 5, written 2 : 5

Key Point

'to' is replaced with ':'

> What is the ratio of black tiles to blue tiles?
> The ratio of black tiles to blue tiles is 5 : 9

Ratios and Fractions

- Ratios can also be written as fractions.

In the previous example:

$\frac{2}{3}$ are women and $\frac{1}{3}$ are men

$\frac{5}{7}$ of the vehicles are cars and $\frac{2}{7}$ of the vehicles are bicycles.

Simplifying Ratios

- The following ratios are equivalent. The relationship between each pair of numbers is the same:

10 : 20
↓
3 : 6
↓
2 : 4
↓
1 : 2   This is a **simpler** way of writing the ratio 10 : 20

- You can simplify a ratio if you can divide by a common factor.
- When a ratio cannot be simplified, it is said to be in its **lowest terms**.

Key Point

When there are no more common factors, the ratio is in its lowest terms.

Simplify the ratio 30 : 100

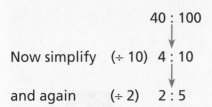 30 ÷ 10   3 : 10   100 ÷ 10   ← Divide both numbers by 10

Write 40p to £1 as a ratio in its lowest terms.

First get the units the same: in pence, £1 is 100p

40 : 100

Now simplify   (÷ 10) 4 : 10

and again   (÷ 2)   2 : 5   ← This is now in its lowest terms.

The angles of a triangle are 20°, 60° and 100°

What is the ratio of the angles in its lowest terms?

20° : 60° : 100°

(÷ 20) 1 : 3 : 5   ← This is now in its lowest terms.

Quick Test

1. Look at this pattern of grey and green tiles:

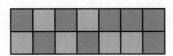

a) Write down the ratio of green tiles to grey tiles.
b) Write down the ratio of grey tiles to green tiles.
2. Write the following ratios in their lowest terms:
a) 3 : 9   b) 28 : 4   c) 25 cm : 1 m

Key Word

lowest terms

Ratio and Proportion 2

You must be able to:

- Share in a given ratio
- Solve problems involving direct and inverse proportion
- Use the unitary method.

Quick Recall Quiz

Sharing Ratios

- Sharing ratios are used when a total amount is to be **shared** or **divided** into a given ratio.

Share £200 in the ratio 5 : 3

Add the ratio to find how many parts there are.

5 + 3 = 8 parts

Divide £200 by 8 to find out how much 1 part is.

200 ÷ 8 = 25

1 part is £25

Now multiply by each part of the ratio.

5 × £25 = £125

3 × £25 = £75

£200 shared in the ratio 5 : 3 is £125 : £75

> **Key Point**
>
> Divide to find one, then multiply to find all.

A sum of money is shared in the ratio 2 : 3

If the smaller share is £30, how much is the sum of money?

2 parts = £30 so 1 part = £30 ÷ 2 = £15

3 parts = £15 × 3 = £45

The sum of money = £30 + £45 = £75

Direct Proportion

- Two quantities are in **direct proportion** if their ratios stay the same as the quantities get larger or smaller.

If the ratio of teachers to students in one class is 1 : 30, then three classes will need 3 : 90

- Graphs of direct proportion are always this shape.
- A straight line is drawn through the points and the graph passes through the origin.
- Using algebra, $s = 30t$, where s is the number of students and t is the number of teachers.

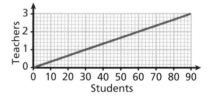

Inverse Proportion

- Two quantities are in **inverse proportion** if as one quantity increases, the other decreases at the same rate.
- Speed and time are in inverse proportion. As speed increases, time decreases.

$$\text{Time} = \frac{\text{Distance}}{\text{Average speed}}$$

Suppose you go on a car journey of 60 km. The time it takes depends on the average speed of the car. The table shows some journey times.

Average speed, s (km/h)	10	20	30	40	50	60
Time, t (h)	6	3	2	1.5	1.2	1

Using algebra, $st = 60$
This is called a **reciprocal graph**.

Graphs of inverse proportion are always this shape. A smooth curve is drawn through the points.

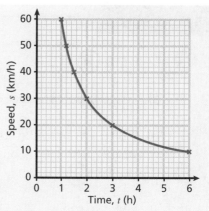

Using the Unitary Method

- Using the unitary method, find the value of **one unit** of the quantity before working out the required amount.

Five loaves cost £4.25. How much will three loaves cost?

One loaf costs £4.25 ÷ 5 = 85 pence

Three loaves will cost 85 pence × 3 = £2.55

Remember: divide to find one, then multiply to find all.

This recipe for making apple pie serves four people:

200 g flour   50 g sugar

200 g butter   8 large apples

Change these amounts to a recipe for 10 people.

Divide to find one, then multiply to find all.

Flour = 200 g ÷ 4 × 10 = 500 g   Sugar = 50 g ÷ 4 × 10 = 125 g

Butter = 200 g ÷ 4 × 10 = 500 g   Large apples = 8 ÷ 4 × 10 = 20

All the amounts have increased in proportion (by $2\frac{1}{2}$ times in this example).

Quick Test

1. Share 40 sweets in the ratio 2 : 5 : 1
2. £360 is divided between Sara and John in the ratio 5 : 4
   How much did each person receive?
3. Work out the missing numbers in these ratios.
   a) 3 : 5 = 12 : ?   b) 4 : 5 = ? : 35
4. If six books cost £30, how much will eight books cost?

Key Words

share
divide
direct proportion
inverse proportion
reciprocal graph

Review Questions

Fractions, Decimals and Percentages

(PS) **1** Convert:

    a) $\frac{3}{20}$ to a percentage [1]

    b) 0.8 to a fraction in its simplest form [2]

(PS) **2** Work out:

    a) 50% of £23 [1]

    b) 10% of 45 cm [1]

    c) 25% of 180 m [1]

    d) 5% of £70 [2]

3 Use the fraction button on your calculator to work out:

    a) $\frac{1}{3}$ of £42 [1]

    b) $\frac{3}{4}$ of £320 [1]

    c) $\frac{4}{5}$ of £850 [1]

Total Marks _____ / 11

(FS) **1** Jenny receives £5 pocket money every week.

    She spends $\frac{1}{2}$ of her money on magazines and $\frac{2}{5}$ on sweets. The rest she saves.

    a) How much does Jenny spend on sweets? [2]

    b) How much does Jenny save? [2]

(PS) **2** A coat costing £90 is reduced by 15% in a sale. What is the sale price of the coat? [3]

(FS) **3** Kim puts £150 into a savings account. She will receive 6% simple interest each year.

    How much will she have in the account after four years? [3]

Total Marks _____ / 10

Equations

(PS) (1) Use inverses to work out the unknown number.

   a) $7 \times \bigcirc = 28$ [1]

   b) $n + 8 = 17$ [1]

   c) $\frac{p}{3} = 6$ [1]

   d) $\bigcirc - 7 = 12$ [1]

(PS) (2) Solve these equations. Show your working.

   a) $4n - 1 = 11$ [2]

   b) $5x + 1 = 21$ [2]

   c) $3a + 8 = 5$ [2]

Total Marks _____ / 10

(PS) (1) Solve these equations. Show your working.

   a) $6x - 5 = 4x + 7$ [2]

   b) $5(x + 2) = 2(x - 1)$ [2]

   c) $3x - 1 = 4 - 2x$ [2]

(FS) (2) Five builders are together paid £20 per hour plus a bonus of £150. They share the pay and each get £50.

Set up and solve an equation to find how many hours they worked. [2]

(3) A vending machine was filled with 56 bars of chocolate at the start of the day.

If 29 are left, how many were sold? Set up and solve an equation to find your answer. [2]

Total Marks _____ / 10

Symmetry and Enlargement

PS **1** Reflect shape A in the dashed mirror lines.

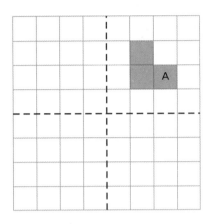

[3]

PS **2** What is the order of rotational symmetry of this shape?

[1]

Total Marks _____ / 4

PS **1** **a)** Rotate shape A 90° clockwise
about the origin (0, 0)
Label the new shape B. [2]

b) Enlarge shape A by a scale factor 3, with a centre
of enlargement (3, 4)
Label the new shape C. [2]

c) Which shapes are congruent? [1]

MR **2** A photograph 5 cm × 7 cm is to be enlarged by a scale
factor of 4

What are the dimensions of the new photograph? [2]

Total Marks _____ / 7

Ratio and Proportion

(PS) **1** What is the ratio of black tiles to white tiles? [1]

(PS) **2** Simplify the following ratios.

a) 8 : 24 [1]

b) 18 : 3 [1]

c) 20 cm : 1 m [2]

d) 80 minutes : 1.5 hours [2]

Total Marks _____ / 7

(FS) **1** Ann and Ben share £450 in the ratio 4 : 5

How much does each person get? [3]

2 A sum of money is shared in the ratio 2 : 3

If the larger share is £27, how much money is there altogether? [3]

(PS) **3** A recipe for six cupcakes needs 40 g of butter and 100 g of flour.

How much butter and flour are needed to make 12 cupcakes? [2]

(PS) **4** A map has a scale of 1 : 50 000

What is the distance on the ground, in km, if a length on the map is:

a) 2.5 cm? [2]

b) 1.4 cm? [2]

Total Marks _____ / 12

Real-Life Graphs and Rates 1

You must be able to:

- Read values from and draw a real-life graph
- Read and draw a conversion graph
- Solve real-life problems using graphs.

Graphs from the Real World

- Graphs from the real world include **conversion graphs**.
- You may also be asked to find conversions without using a graph.
- You may have to convert between:
  - pounds (£) and US dollars ($)
  - pounds (£) and euros (€)
  - pints and litres
  - mph and km/h
  - miles and kilometres
  - gallons and litres

Reading a Conversion Graph

- To convert from one unit to the other, read straight across to the line, then go straight down until you reach the other axis. To convert the other way, go up until you reach the line, then read across.

Convert 30 miles into kilometres.

Draw a line **straight up** from 30 miles until it hits the line.

Go **straight across** to the kilometres axis.

30 miles is **equivalent** to 48 km.

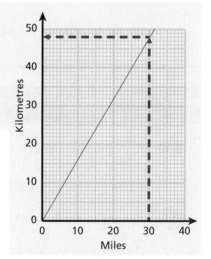

Drawing a Conversion Graph

- To find the points you need to plot, work out a number of equivalent values. Join the plotted points with a straight line.

Jack's company pays him 80 pence for each mile he travels. Use the information to draw a graph of his pay.

Distance in miles	0	10	20	30
Amount	0	£8	£16	£24

Work out how much Jack will be paid for different journeys.

Use your table to plot at least three points, and join them with a straight line.

Key Point

Extend your line to the edge of the graph grid.

The graph can now be read to find the pay for different journeys.

Solving Real-Life Problems Using Graphs

- You can use graphs to solve problems set in real-life situations.

The graph shows the charge to hire a minibus. There is a fixed charge for any distance up to 2 miles of £10 and then the charge is £3 per extra mile.

Work out the cost of hiring the minibus for a journey of 30 miles.

The first 2 miles will cost £10
The next 28 miles will cost 28 × £3 = £84
£10 + £84 = £94

It will cost £94 to hire the minibus for a 30-mile journey.

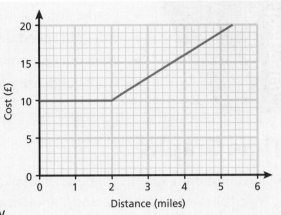

This **exponential graph** shows the number of infections of a disease doubling each day. How does the rate of infections change over four days?

On day 0 there are 2000 infections; day 1 has 4000; day 2 has 8000; day 3 has 16000; and day 4 has 32000
So over four days, infections have risen from 2000 to 32000

This is an increase by a factor of 16 (32000 ÷ 2000)

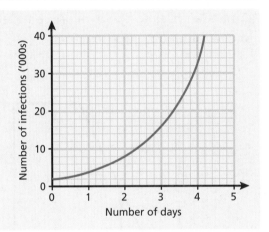

Quick Test

Use the conversion graph on page 102 for questions 1 and 2.
1. Find how many kilometres are equivalent to:
   a) 25 miles   b) 10 miles
2. How many miles are equivalent to:
   a) 30 km?   b) 40 km?
3. Sharon charges £1 for the use of her taxi and 50p per mile after that. Work out the cost of a journey that is:
   a) 4 miles
   b) 6 miles
   c) Use the information to draw a graph of her charges.

Key Words

conversion graph
exponential graph

Real-Life Graphs and Rates 2

Quick Recall Quiz

You must be able to:

- Read a distance–time graph
- Work out speed, distance and time
- Work out unit prices
- Work out density, mass and volume.

Time Graphs

- Distance–time graphs give information about journeys. Use the horizontal scale for time and the vertical scale for **distance**.
- Distance–time graphs are also used to calculate **speed**.

Amanda cycles to the gym and back every Sunday. The graph below shows Amanda's journey.

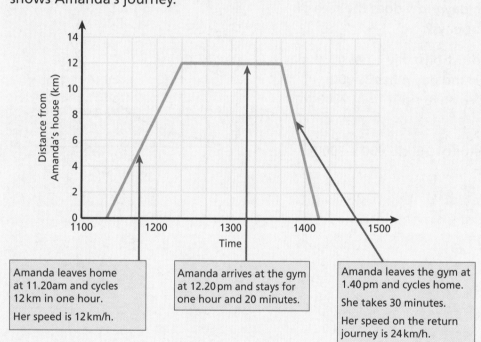

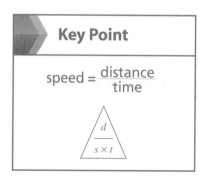

Amanda leaves home at 11.20am and cycles 12 km in one hour.

Her speed is 12 km/h.

Amanda arrives at the gym at 12.20 pm and stays for one hour and 20 minutes.

Amanda leaves the gym at 1.40 pm and cycles home.

She takes 30 minutes.

Her speed on the return journey is 24 km/h.

Travelling at a Constant Speed

- When the speed you are travelling at does not change, it remains **constant**.
- You can work out speed, distance or time using a formula triangle.

A car travels 120 miles at 40 miles per hour.

How long does the journey take?

time = distance ÷ speed

time = 120 ÷ 40 = 3 hours

Cover up what you are trying to find.

Key Point

$$\text{speed} = \frac{\text{distance}}{\text{time}}$$

A plane takes $2\frac{1}{2}$ hours to travel 750 miles. What is the speed of the plane?

speed = distance ÷ time

speed = 750 ÷ 2.5

= 300 mph

Cover up what you are trying to find.

Unit Pricing

- Unit pricing involves using what 'one' is to work out other amounts.

If £1 = \$1.75, how much would a pair of jeans cost in \$ if they were £60?

$$60 \times 1.75 = \$105$$

How much would a TV costing \$525 be in £?

$$525 \div 1.75 = £300$$

Multiply to get the dollars; divide to get the pounds.

A 250-gram bag of pasta costs £1.25

a) Work out the cost per gram.

b) Work out the number of grams bought for 1p.

a) Cost per gram = 125p ÷ 250 = 0.5p

b) Number of grams bought for 1p = 2 grams

Density

- You can work out **density**, mass and volume using a formula triangle similar to speed.

density = mass ÷ volume

Find the density of an object that has a mass of 60 g and a volume of 25 cm³

density = mass ÷ volume

= 60 ÷ 25

= 2.4 g/cm³

Cover up what you are trying to find.

Key Point

Remember to use the correct units.

Volume: **cm³** and **m³**

Mass: **g** and **kg**

Density: **g/cm³** and **kg/m³**

Quick Test

1. What is 90 minutes in hours?
2. Stuart drives 180 km in 2 hours 15 minutes. Work out his average speed.
3. John travelled 30 km in $1\frac{1}{2}$ hours. Kamala travelled 42 km in 2 hours. Who had the greater average speed?
4. If £1 = €1.20, what would £200 be worth in €?
5. What is the mass of 250 ml of water with density of 1 g/cm³? 1000 cm³ = 1 litre

Key Words

distance

speed

density

Right-Angled Triangles 1

You must be able to:

- Label right-angled triangles correctly
- Understand Pythagoras' Theorem
- Find the length of the longest side
- Find the length of a shorter side.

Pythagoras' Theorem

- Remember the formula for **Pythagoras' Theorem**: $a^2 + b^2 = c^2$

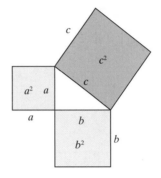

> ### Key Point
>
> The longest side, the hypotenuse, is called c and is opposite the right angle.
>
> The two shorter sides are called a and b. The order is not important.

Finding the Longest Side

- To find the longest side (**hypotenuse**), **add** the **squares**. Then take the **square root** of your answer.

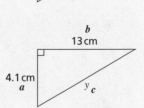

Find the length of y. Give your answer to 1 decimal place.

First label the sides a, b and c.

Now use the formula:

$$a^2 + b^2 = c^2$$
$$4.1^2 + 13^2 = y^2$$
$$16.81 + 169 = y^2$$
$$185.81 = y^2$$
$$y = \sqrt{185.81}$$
$$= 13.6\,\text{cm} \ (1 \ \text{d.p.})$$

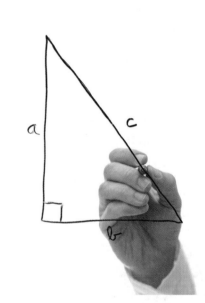

Finding a Shorter Side

- To find a shorter side, **subtract** the squares. Then take the square root of your answer.

Find the length of y. Give your answer to 1 decimal place.

 Label the sides a, b and c.

Now use the formula:

$a^2 + b^2 = c^2$

$7^2 + y^2 = 14^2$

$49 + y^2 = 196$

$y^2 = 196 - 49 = 147$

$y = \sqrt{147} = 12.1\,\text{cm}$ (1 d.p.)

Find the length of y. Give your answer to 1 decimal place.

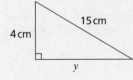 Label the sides a, b and c.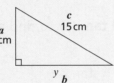

Now use the formula:

$a^2 + b^2 = c^2$

$4^2 + y^2 = 15^2$

$16 + y^2 = 225$

$y^2 = 225 - 16 = 209$

$y = \sqrt{209} = 14.5\,\text{cm}$ (1 d.p.)

Key Point

You will always $\sqrt{}$ at the end.

Quick Test

1. Work out:
   a) 3.2^2
   b) 15.65^2
2. Work out:
   a) $\sqrt{4900}$
   b) $\sqrt{39.69}$
3. Work out the longest side of a right-angled triangle if the shorter sides are 5 cm and 2.2 cm.
4. Work out the shorter side of a right-angled triangle if the longest side is 12 cm and the other shorter side is 9 cm.

Key Words

Pythagoras' Theorem
hypotenuse
square

Right-Angled Triangles 2

You must be able to:

- Remember the three ratios
- Work out the size of an angle
- Work out the length of an unknown side.

Side Ratios

- Label the sides of the triangle in relation to the angle that is marked.

H is always the longest side

Hypotenuse (H)

Opposite (O)
opposite the angle

x

Adjacent (A)
next to the angle

- There are three ratios: **sin**, **cos** and **tan**. Try to find a way of remembering these:

> Sin is short for 'sine', cos is short for 'cosine' and 'tan' is short for 'tangent'.

$$\sin x = \frac{O}{H} \qquad \cos x = \frac{A}{H} \qquad \tan x = \frac{O}{A}$$

- You can use the formula triangles:

- You can use a rhyme:
  Some **O**ld **H**orses **C**an **A**lways **H**ear **T**heir **O**wners **A**pproach

Use your calculator to work out the ratios for these angles:

a) $\sin 60° = 0.8660$
b) $\cos 45° = 0.7071$
c) $\tan 87° = 19.0811$

> **Key Point**
>
> Ensure your calculator is in '**degree**' mode.

Use your calculator to work out the angles for these ratios:

a) $\sin x = 0.5$ $\qquad\qquad x = \sin^{-1} 0.5 = 30°$

b) $\cos x = \frac{3}{5}$ $\qquad\qquad x = \cos^{-1} (3 \div 5) = 53.1°$

c) $\tan x = 2.9$ $\qquad\qquad x = \tan^{-1} 2.9 = 71°$

Finding Angles in Right-Angled Triangles

- You need to know two sides of the triangle to find an angle.

Find the size of angle x. Give your answer to 1 decimal place.

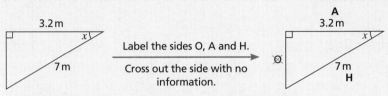

Label the sides O, A and H.

Cross out the side with no information.

As **A** and **H** are known, we use the cos ratio.

$\cos x = \dfrac{A}{H}$ so $\cos x = \dfrac{3.2}{7}$

$x = \cos^{-1}\left(\dfrac{3.2}{7}\right)$

$x = 62.8°$

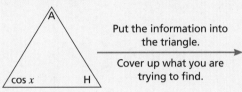

Put the information into the triangle.

Cover up what you are trying to find.

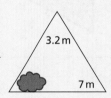

Finding the Length of a Side

- You need to know the length of one side and an angle to find the length of another side.

Find the length of the side labelled y. Give your answer to 1 decimal place.

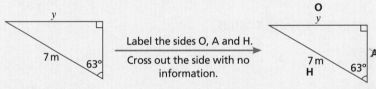

Label the sides O, A and H.

Cross out the side with no information.

As **O** is to be found and **H** is known, we use the sine ratio.

$\sin x = \dfrac{O}{H}$ so $\sin 63° = \dfrac{y}{7}$

$7 \times \sin 63° = y$

$y = 6.2$ m

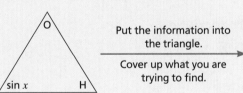

Put the information into the triangle.

Cover up what you are trying to find.

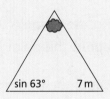

Quick Test

1. Use your calculator to work out the ratios for these angles.
   a) $\sin 20°$　　　b) $\cos 30°$　　　c) $\tan 45°$
2. Use your calculator to work out the angles for these ratios.
   a) $\sin x = 0.8337$　　　b) $\cos x = \dfrac{4}{7}$　　　c) $\tan x = 32$

Key Words

sin
cos
tan

Review Questions

Symmetry and Enlargement

(PS) **1** Reflect shape A in the dashed mirror line.

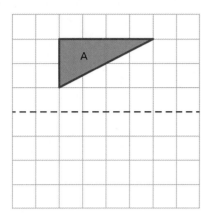

[2]

Total Marks _____ / 2

(PS) **1** What is the order of rotational symmetry of these shapes? [3]

a)   b)   c)

(PS) **2** **a)** Rotate shape A 90° clockwise about the point (1, 0)

Label the new shape B. [2]

b) Enlarge shape A by a scale factor 2 from the point (–3, 3)

Label the new shape C. [2]

c) Which shapes are congruent? [1]

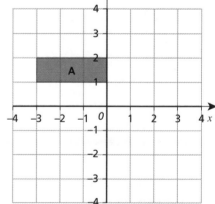

(MR) **3** A house is 6 m tall in real life. The same house has been drawn on paper as being 3 cm tall.

What is the scale in the form 1 : n? [2]

Total Marks _____ / 10

Ratio and Proportion

(PS) **1** There are 14 boys and 16 girls in a class.

What is the ratio of girls to boys? Write your answer in its lowest terms. [2]

(PS) **2** Simplify the following ratios.

a) 10 : 2 [1]

b) 16 : 24 [1]

c) 25 pence : £2 [2]

3 For each ratio table, work out the missing value.

a)

grams	%
4	100
8	

[1]

b)

£	%
20	100
	50

[1]

c)

km	%
16	100
	125

[1]

Total Marks .... / 9

1 Complete the equivalent ratios.

a) 8 : 3 = ? : 15 [1]

b) 7 : ? = 63 : 108 [1]

(FS) **2** A sum of money is shared in the ratio 1 : 4

If the larger share is £120, how much money is there altogether? Show your working. [3]

3 Share 40 pens in the ratio 3 : 5, showing your working. [2]

(FS) **4** Danesh bought 18 postcards for £2.16

How much would he pay if he bought 27 postcards? Show your working. [2]

(PS) **5** A telegraph pole 60 feet high casts a shadow 12 feet long. At the same time of day, how long is the shadow cast by a 90-foot pole? [2]

Total Marks .... / 11

Practice Questions

Real-Life Graphs and Rates

1 Concorde could travel 20 miles every minute.

How many miles per hour (mph) is that? [2]

2 Use £1 = €1.19 to work out how much £3.50 is in euros. [2]

Total Marks .... / 4

1 The table shows the prize money for the winners of professional tennis tournaments in Australia and France in one particular year.

Country	Money
Australia	1 000 000 Australian dollars (£1 = 2.70 Australian dollars)
France	780 000 euros (£1 = 1.54 euros)

Which country paid more money? You must show your working. [2]

(PS) 2 The graph shows the flight details of an aeroplane travelling from London to Madrid, via Brussels.

a) What is the aeroplane's average speed from London to Brussels? [2]

b) At what time did the plane arrive in Madrid? [1]

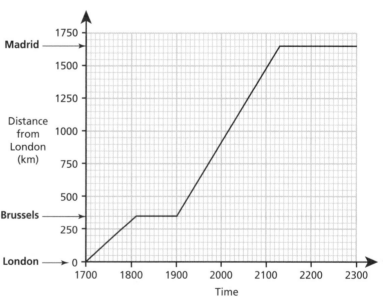

Total Marks .... / 5

Right-Angled Triangles

PS **1** Use your calculator to work out the value of:

a) 17^2 [1]

b) 3.5^2 [1]

c) $\sqrt{529}$ [1]

d) $\sqrt{40.96}$ [1]

Total Marks _____ / 4

PS **1** Use Pythagoras' Theorem to work out:

a) the length of AC.

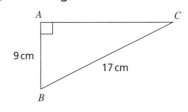

b) the length of BC.

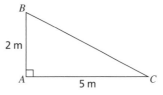

[4]

PS **2** a) Work out the value of P.

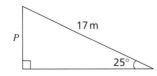

b) Work out the size of angle y.

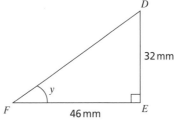

[4]

3 Work out the size of angle x.

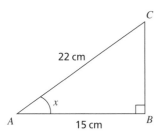

[3]

Total Marks _____ / 11

Review Questions

Real-Life Graphs and Rates

(FS) **1** Use £1 = US$1.75 to work out how much:

    a) $200 is in £ [2]

    b) £200 is in US$ [2]

(PS) **2** A coach travels 300 miles non-stop at an average speed of 60 mph.

    a) For how many hours does the coach travel? [2]

    b) At the same speed, how far will the coach travel in four hours? [2]

Total Marks _____ / 8

(PS) **1** Calculate the density of a piece of metal that has a mass of 2000 kg and a volume of 5 m³ [2]

2 The graphs show information about the different journeys of four people.

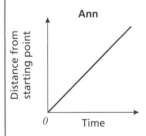

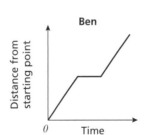

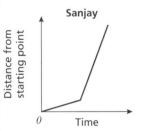

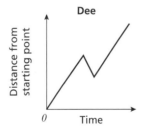

Name	Journey description
	This person walked slowly and then ran at a constant speed.
	This person walked at a constant speed but turned back for a while before continuing.
	This person walked at a constant speed without stopping or turning back.
	This person walked at a constant speed but stopped for a while in the middle.

Copy the table and write the correct names next to the journey descriptions. [2]

Total Marks _____ / 4

Right-Angled Triangles

(PS) **1** Use your calculator to work out the value of:

    a) 3.3^2 [1]

    b) $\sqrt{196}$ [1]

(PS) **2** Use Pythagoras' Theorem to work out:

    a) the length of y.   **b)** the length of x.

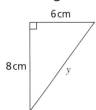

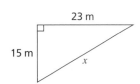

[4]

Total Marks _____ / 6

(PS) **1**   **a)** Work out the size of angle x.   **b)** Work out the length of AB.

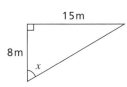

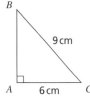

[4]

2 An isosceles triangle has a base of length 4 cm and a perpendicular height of 8 cm.

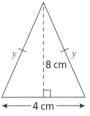

Giving your answers correct to 1 decimal place, calculate:

    a) the length, y, of one of the equal sides. [3]

    b) the perimeter of the triangle. [1]

Total Marks _____ / 8

Mixed Test-Style Questions

No Calculator Allowed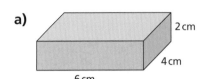

1   Work out both the surface area and volume of these cuboids.

a)

2 cm
4 cm
6 cm

Surface area = _____ cm²

Volume = _____ cm³

b)

7 cm
12 cm
8 cm

Surface area = _____ cm²

Volume = _____ cm³

4 marks

2   Solve the following, giving your answers in the simplest form.

a) $4\frac{1}{2} + 2\frac{1}{3}$

b) $5\frac{2}{3} + 8\frac{1}{4}$

c) $9\frac{1}{6} - 2\frac{3}{8}$

d) $12\frac{1}{2} - 14\frac{5}{6}$

4 marks

3 On the grid below draw the lines for:

a) $y = 6$ **b)** $x = -4$ **c)** $y = x$

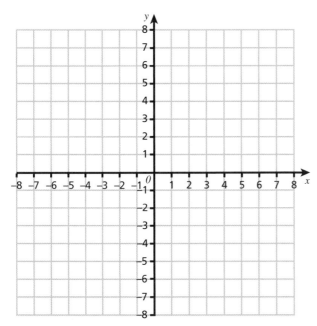

3 marks

4 **a)** Complete the table of values for the equation $y = 3x - 3$

x	−2	−1	0	1	2	3
y						

b) Plot the coordinates on the graph below and join them with a line.

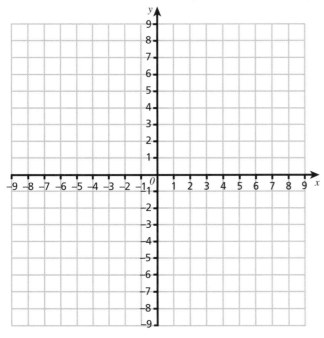

4 marks

TOTAL

15

Mixed Test-Style Questions

5 Calculate angles x and y.

a)

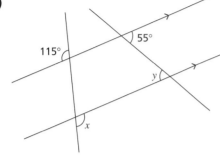

$x =$ .... $^\circ$

$y =$ .... $^\circ$

b)

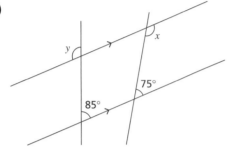

$x =$ .... $^\circ$

$y =$ .... $^\circ$

4 marks

6 Simplify the following expressions:

a) $3x - 2y + x + 6y$

b) $4g + 5 - g - 4$

2 marks

7 Expand the following expressions:

a) $4(x - 5)$

b) $4x(x + 4)$

2 marks

8 Factorise completely the following expressions:

a) $6x - 12$

b) $4x^2 - 8x$

2 marks

9 The rectangle and trapezium below have the same area.

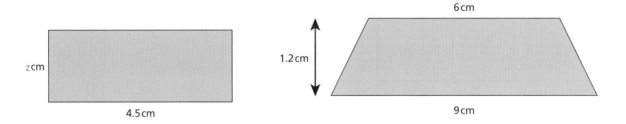

6 cm

1.2 cm

z cm

4.5 cm

9 cm

Work out the value of z. Show your working.

3 marks

10 A chocolate bar costs 60p. A vending machine which sells the chocolate bars is emptied and the following coins are found:

Coins	Frequency
£1	25
50p	59
20p	72
10p	31

How many chocolate bars were sold?

3 marks

TOTAL

16

11 Use the cards below to make two-digit numbers as asked. The first one has been done for you.

| 7 | 1 | 6 | 5 | 4 |

An odd number | 7 | 1 |

a) A square number

b) A multiple of 3

c) A prime number

2 marks

12 A plant grows by 10% of its height each day. At 8 am on Monday the plant was 400 mm high.

How tall was it:

a) at 8 am on Tuesday?

b) at 8 am on Wednesday?

2 marks

13 What are the names of the following shapes?

a)

....

b)

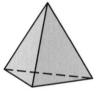

....

c)

....

3 marks

14

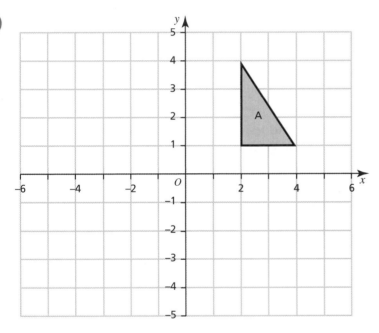

a) Reflect shape A in the *y*-axis.

b) Enlarge shape A by a scale factor 2 from the point (3, 4)

c) Rotate shape A 180° about (0, 0)

3 marks

TOTAL

10

Calculator Allowed

1 Sam was sitting on the dock of the bay watching boats for an hour. He collected the following information:

Type	Frequency	Probability
Tug boat	12	
Ferry boat	2	
Sail boat	16	
Speed boat	10	

a) Complete the probability column in the table, giving your answers as fractions.

b) If Sam saw 75 boats, estimate how many of them would be sail boats.

5 marks

2 Work out the surface area and volume of these cylinders.

a) radius = 4 cm

9 cm

Surface area = _____ cm²

Volume = _____ cm³

b) diameter = 10 cm

4.5 cm

Surface area = _____ cm²

Volume = _____ cm³

8 marks

3 The diagram shows a circle inside a square of side length 4 cm.

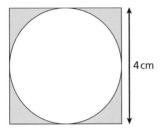

4 cm

Find the total area of the shaded regions.

.... cm²

3 marks

4 Barry is planning to buy a car. He visits two garages which have the following payment options:

Mike's Motors	Carol's Cars
£500 deposit	£600 deposit
36 monthly payments of £150	12 monthly payments of £50
£150 administration fee	24 monthly payments of £200

Which garage should Barry buy his car from in order to get the cheaper deal?
Show working to justify your answer.

3 marks

TOTAL

19

5 An athlete can run 100 m in 12 seconds.

Work out the athlete's speed in:

a) m/s

....m/s

b) km/h

....km/h   ☐ 4 marks

6 A wire 15 m long runs from the top of a pole to the ground as shown in the diagram. The wire makes an angle of 45° with the ground.

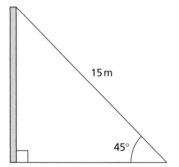

Calculate the height of the pole. Give your answer to a suitable degree of accuracy.

☐ 2 marks

7 Below is a map of an island. The scale is 1 cm : 4 km

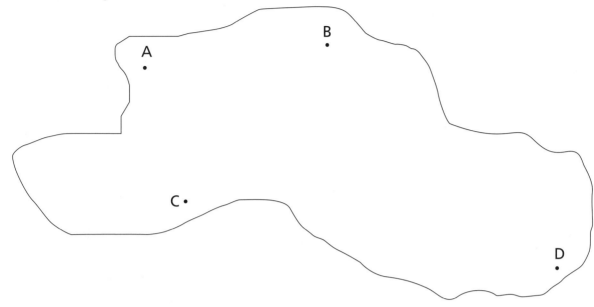

A helicopter flies directly from A to B, B to C, then C to D.

What is the total distance flown in kilometres?

.................................km

3 marks

8 Change 8% to:

a) a decimal

b) a fraction in its simplest form

2 marks

TOTAL

11

Mixed Test-Style Questions

9 A recipe for 12 cupcakes needs 80 g of butter and 200 g of flour.

How much butter and flour are needed to make:

a) 24 cupcakes?

....g of butter

....g of flour

b) 30 cupcakes?

....g of butter

....g of flour

4 marks

10 Work out the missing numbers. You can use the first line to help you.

$16 \times 21 = 336$

a) $16 \times$ .... $= 112$

b) $336 \div 21 =$ ....

2 marks

11 The data below represents the waiting time in minutes of 15 patients in a doctor's surgery:

49	23	34	10	28	28	25	28
39	35	15	14	48	10	20	

a) Work out the modal waiting time.

b) Work out the median waiting time.

c) Work out the range of the waiting times.

4 marks

TOTAL

10

Answers

Pages 6–7 **Review Questions**

Page 6

1. 1996 **[1]** because 2000 – 1996 = 4 but 2007 – 2000 = 7 **[1]**
2.

	4	7	6	245	**[2]**
–	2	3	1	**OR** Method mark for valid	
	2	4	5	attempt to subtract	**[1]**

3. 5000, 46 000, 458 000, 46 000   All 4 correct **[2]**; Any 3 correct **[1]**
4. $\frac{27}{50}$, 55%, 0.56, 0.6, 0.63
   All 5 correct **[3] OR** 3 out of 5 correct **[2] OR** 55% = 0.55 and
   $\frac{27}{50}$ = 0.54 **[1]**
5. Ahmed = 2 × 22 = 44 years old **[2]**
   OR Rebecca = 25 – 3 = 22 years old seen **[1]**
6. 15 878 **[3]**
   OR 12 000 + 1800 + 210 + 1600 + 240 + 28 **[2]**
   OR Valid attempt at multiplication with one numerical error **[1]**
7. 52 **[2]**
   OR Valid attempt at division with one numerical error **[1]**
8.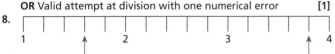
   Both correct **[2]**; One correct **[1]**

Page 7

1. 200 ml **[3]**
   OR 1000 ml and 800 ml seen **[2]**
   OR 1000 ml seen **[1]**
2. a) 16 **OR** 36 **[1]**
   b) 13 or 17 or 31 or 37 or 53 or 61 or 67 or 71 or 73 **[1]**
   c) 36 **[1]**
   d) 15 **[1]**
3. 9 cm **[2] OR** 27 ÷ 3 seen **[1]**

   An equilateral triangle has three equal sides.

4. 33 **[2]**
   OR 11 seen **[1]**
5. 25° **[3]**
   OR 65 seen **[2]**
   OR 130 seen **[1]**

   Angles in a triangle add up to 180°. Base angles in an
   isosceles triangle are equal. A right angle is 90°.

6. $T = 200$ **[1]** ($S = T + 100$, $2T + 100 = 500$) $S = 300$ **[1]**

Pages 8–15 **Revise Questions**

Page 9 Quick Test

1. 3601, 3654, 3750, 3753, 3813
2. 226 635
3. 163
4. 40
5. −5 < 3

Page 11 Quick Test

1. a) 49   b) 64
2. a) 7   b) 3
3. $2^3 \times 5$
4. 252
5. 8

Page 13 Quick Test

1. Output 4; Input 48
2. 21, 25, 29, 33, 37
3. ÷2 **OR** × $\frac{1}{2}$
4. B
5. 39

Page 15 Quick Test

1. 8, 13, 18, 23, 28
2. 5, 8, 13, 20, 29
3. a) $3n + 3$ **OR** $3(n + 1)$
   b) 153
4. Position-to-term rule

Pages 16–17 **Practice Questions**

Page 16

1. a) January **[1]**
   b) July **[1]**
   c) 12 **[1]**
2. $\sqrt{49} = 7$ cm

1. Jessa is right **[1]**
   Using BIDMAS multiply is first so 3 + 20 + 7 = 30 **[1]**
2. a) £4248 **[3]**
   OR 3000 + 500 + 40 + 600 +100 + 8 **[2]**
   OR Valid attempt at multiplication with one numerical error **[1]**
   b) 354 ÷ 52 = 6.8 **[2]**
   OR Valid attempt at division with one numerical error **[1]**
   7 coaches needed **[1]**
   c) 10 spare seats **[2]**
   OR 364 seen   (52 × 7 = 364) **[1]**
3. $8^2 = 64$ and $9^2 = 81$ so $\sqrt{79}$ is between 8 and 9 **[2]**
   OR attempt to find any two square numbers each side of 79 **[1]**

Page 17

1. a) 19 **[1]** 23 **[1]**
   b) + 4 **[1]**
2. a) 14 **[1]**
   b) 9 weeks **[1]**

1. a) $3n + 1$ **[3]**
   OR $3n$ **[2]**
   OR +3 seen as term-to-term rule **[1]**
   b) 181 **[1]**
2. 5, 9, 13, 17, 21 = arithmetic
   2, 8, 18, 32, 50 = quadratic
   8, 17, 32, 53, 80 = quadratric
   All three correct **[2] OR** one correct **[1]**

Pages 18–25 **Revise Questions**

Page 19 Quick Test

1. 24 cm
2. 27 cm²
3. 6 cm²
4. Area 19 cm²; Perimeter 24 cm

Page 21 Quick Test

1. 16 cm²
2. 13 cm²
3. Circumference = 37.7 cm (1 d.p.); Area = 113.1 cm² (1 d.p.)
4. Circumference = 12.6 cm (1 d.p.); Area = 12.6 cm² (1 d.p.)

Page 23 Quick Test

1. a) Red = 9; Blue = 7; Green = 8; Yellow = 6; Other = 7
   b) 37
2. a) 6 and 3
   b) Mean = 9.2 (1 d.p.); Median = 6; Range = 37
   c) Median as there is an outlier

Page 25 Quick Test

1. a)

	Football	Rugby	Total
Women	16	9	25
Men	20	10	30
Total	36	19	55

   b) 36   c) 9   d) 55

Pages 26–27 Review Questions

Page 26

1. £550 **[3] OR** £2200 ÷ 4 seen **[2] OR** £2200 seen **[1]**
2. Mercury, Venus, Earth, Mars, Jupiter, Saturn, Uranus, Neptune
   All 8 correct **[2] OR** 6 correct **[1]**
3. 25 cm²   **[2]**
   OR 5 cm seen   **[1]**
4. 64   **[2]**
   OR 4×4 or 6×8 seen   **[1]**

1. 1248 **[1]** (24 is half of 48) 26 **[1]** (26 is half of 52) 48 **[1]**
2. 15 and 12 **[2] OR** either 15 **OR** 12 seen **[1]**
3. 4 packs of sausages and 3 packs of rolls   **[3]**
   OR 24 seen   **[2]**
   OR Valid attempt to find LCM of 6 and 8 seen   **[1]**

Page 27

1. a) 9 **[1]** 16 **[1]**
   b) **[1]** for each correct answer e.g. $n \div 6$ (or equivalent) or $\sqrt{n}$

1. a) 44 **[1]**   For 20th term $n = 20$
   b) 204 **[1]**
   c) $2n - 1$ **[2] OR** $2n$ seen **[1]**
2. 4.03pm **[3] OR** 180 seen **[2] OR** Valid attempt to split 20 and
   45 into prime factors seen **[1]**

   This question is asking for the LCM.

Pages 28–29 Practice Questions

Page 28

1. Perimeter = 20 cm **[1]**; Area = 24 cm² **[1]**
2. 32 cm² **[2] OR** 8×4 seen **[1]**

1. $x = 8$ cm **[1]** $y = 6.8$ cm **[1]**   Area of a rectangle = $l \times w$

2. a) 192 **[3] OR** 16×12 seen **[2] OR** 120 000 and 625 seen **[1]**

   As each tile is 25 cm, 16 will fit along one side and 12 along the other.

   b) £300 **[2] OR** 20 seen **[1]**   The nearest multiple of 10 bigger than 192 is 200

   c) 8 tiles   **[1]**

Page 29

1. a) $(6 + 11 + 9 + 12 + 7) \div 5 = 9$ **[1]**
   b) 6, 7, 9, 11, 12 numbers in order; 9 is in the middle   **[1]**
   c) Any five numbers with a mean of 9   **[1]**
2. 20 **[2] OR** 90 seen **[1]**
3. a) $16 + 5 + 4 + 2 = 27$ **[1]**
   b) 20   **[1]**
   c) $2 + 6 + 7 + 15 + 11 = 41$ **[1]**

1. a) Phil 66.4 (1 d.p.) **[1]**; Dave 68.9 (1 d.p.) **[1]**
   b) Phil 70 **[1]** Dave 175 **[1]**
   c) Dave as higher average **OR** Phil as more consistent   **[2]**

Pages 30–37 Revise Questions

Page 31 Quick Test

1. a) 235.6   b) 5.6781

2. 100 000
3. 16.2, 16.309, 16.34, 16.705, 16.713

Page 33 Quick Test
1. 49.491
2. 17.211
3. 163
4. 150 000

Page 35 Quick Test
1. $7x + 5y + 6$
2. c^2d^2
3. 14

Page 37 Quick Test
1. $8x - 4$
2. $2x - 14y$
3. $5(x - 5)$
4. $2x(x - 2)$

Pages 38–39 Review Questions

Page 38

1. $\pi \times 49$ **[1]** = 153.9 cm² **[1]**
2. Area = 45 cm² **[1]**; Perimeter = 28 cm **[1]**
3. Area = $\frac{1}{2}(6 + 8) \times 4$ **[1]** = 28 cm² **[1]**

1.
   This is a compound area separated into a rectangle and triangle.

   a) 22 m² **[3] OR** 20 and 2 seen **[2] OR** only 20 seen **[1]**
   b) 4 tins   **[1]**
   c) £48 **[2] OR** answer to b) × 12 seen **[1]**
2. Any answer from 5092 to 5095 inclusive (depending on value of pi used) **[2] OR** 157.0796 or 1.57 seen **[1]**

   This question is about the circumference of a circle; notice the units are different, 50 cm = 0.5 m

Page 39

1.

	Boys	Girls
Right-handed	12	14
Left-handed	7	1

   Completely correct **[3] OR** 3 boxes correct **[2] OR** 2 boxes correct **[1]**

2. a)

Vegetable	Frequency
Carrots	4
Peas	8
Potatoes	9
Sweetcorn	3

   All correct **[2] OR** 3 rows correct **[1]**
   b) Potatoes   **[1]**

1. a) Median **[1]** as data contains an outlier **[1]** (14 808 much bigger than the rest of the data)
   b) 14 808   Mode is the most common.   **[1]**

Pages 40–41 Practice Questions

Page 40

1. $3.7 \times 10 \rightarrow 37$; $3.7 \times 100 \rightarrow 370$; $3.7 \div 10 \rightarrow 0.37$; $3.7 \div 100 \rightarrow 0.037$;
   $3.7 \times 1000 \rightarrow 3700$. All correct **[3] OR** 2 correct **[2] OR** 1 correct **[1]**
2. 0.5679 greater **[1]** Has 6 hundredths compared to 5 **[1]**

1. a) 57.832   **[1]**   b) 21.98   **[1]**
   c) 74.154   **[1]**   d) 216   **[1]**
2. 16.061, 16.94, 17.09, 17.203, 17.84   **[2]**
   [1] for any four in correct order, ignoring fifth value.

Page 41

1. $5x = 25$, $x = 9$, $4x = 28$. All correct **[2]** OR 2 correct **[1]**
2. $7y$ **[1]** $12y$ **[1]**

1. Both of them **[1]** $2(x + y)$ expands to $2x + 2y$ OR vice versa **[1]**
2. £123 **[3]** OR 48 seen **[2]** OR 120×0.4 seen **[1]**
3. $3x + 7$ **[2]** OR $8x + 2 - 5x + 5$ seen **[1]**
4. $3a(bc + 2)$ **[2]** OR $3(abc + 2a)$ OR $a(3bc + 6)$ seen **[1]**

   'Completely' means remove all common factors.

5. 144 **[2]** OR 9 seen **[1]**

   $ab = a \times b$

6. $3a - b$, $2a - b$, b. All three correct **[2]**; any one correct **[1]**

Pages 42–49 Revise Questions

Page 43 Quick Test
1. Volume = 140 cm³
   Surface area = 166 cm²
2. Volume = 440 cm³
   Surface area = 358 cm²

Page 45 Quick Test
1. **a)** Volume = 197.9 cm³
      Surface area = 188.5 cm²
   b) Volume = 603.2 cm³
      Surface area = 402.1 cm²
2. 174 cm³

Page 47 Quick Test
1. 20°
2. Sunday
3. No, Helen still used more (approx. mean 51).

Page 49 Quick Test
1. e.g. money spent on advertising against sales for that item
2. Own goal

Pages 50–51 Review Questions

Page 50

1. 0.759 **[2]** OR 253×3 seen **[1]**
2. 17 **[2]** OR $85 \div 5$ seen **[1]**
3. 5.2 and 3.8 circled **[1]**
4. 140 **[2]** OR 7000 and 50 seen **[1]**
5. £854.40 **[2]** OR Valid attempt at multiplication with one numerical error **[1]**

1. 6.765, 6.776, 7.675, 7.756, 7.765: all 5 correct **[2]** OR 3 correct **[1]**
2. **a)** £15.99 **[2]** OR attempt to add up three costs with only one numerical error **[1]**
   b) £4.01 **[1]**
3. 6.93 **[1]**
4. $-0.5 \leqslant \text{error} < 0.5$ **[1]**

Page 51

1. **a)** $11k + 5$ **[1]**
   b) $4k + 1$ **[1]**
2. $-2k$ **[1]**
3. cd^2, c^2d, c^2d^2. All three correct **[2]** OR any two correct **[1]**

1. **a)** ab **[1]**
   b) $2a + 2b$ or $2(a + b)$ **[1]**
   c) $3a$ and $5a$ **[1]**
2. $4t(2ut - u + 5)$ **[2]**
3. 3600 metres per hour **[2]** $1200 \div \frac{1}{3}$ OR 3 seen **[1]**

Page 52

1.

Cuboid	6	12	8	**[1]**
Square-based pyramid	5	8	5	**[1]**
Hexagonal prism	8	18	12	**[1]**

2. **a)** Cylinder **[1]** **b)** Triangular prism **[1]** **c)** Cube **[1]**
3. **a)** 150 cm³ **[1]**; 190 cm² **[1]**
   b) 81 cm³ **[1]**; 117 cm² **[1]**

1. $942 \div 5^2 \div \pi$ **[1]** = 12.0 cm **[1]**
2. $1385 \div 7^2 \div \pi$ **[1]** = 9.00 cm **[1]**

Page 53

1. $\frac{360}{45} = 8$

Category	Frequency	Angle	
Brie	21	$21 \times 8 = 168°$	**[1]**
Cheddar	5	$5 \times 8 = 40°$	**[1]**
Stilton	14	$14 \times 8 = 112°$	**[1]**
Other	5	$5 \times 8 = 40°$	**[1]**
Total	45	360°	

Favourite Cheese

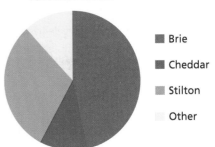

- Brie
- Cheddar
- Stilton
- Other

All angles of pie chart correct to within 2° **[1]**; correct labelling **[1]**
2. $20 - 8 = 12$ **[1]**

1. Example answer: Fuel remaining **[1]** against distance travelled **[1]**
2. How many more times do you go shopping during the Christmas period than other times of the year?

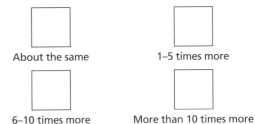

About the same   1–5 times more

6–10 times more   More than 10 times more

Question should have a time frame (e.g. How many more times do you go shopping during the Christmas period than other times of the year?). **[2]**
Response boxes should cover all outcomes and not overlap (e.g. about the same; 1–5 times more; 6–10 times more; more than 10 times more). **[2]**
3. **a)** Scatter graph, frequency polygon, pie chart, bar chart, line graph, histogram, etc.   Four examples **[2]**
   b) Choose a graph with a reason, for instance pie chart as it can show the proportions of different amounts of time spent. **[2]**

Page 55 Quick Test

1. e.g. $\frac{4}{6}\frac{6}{9}\frac{8}{12}\frac{10}{15}\frac{12}{18}$

2. $\frac{64}{77}$

3. $\frac{29}{72}$

4. $\frac{15}{52}$

5. $\frac{81}{100}$

Page 57 Quick Test

1. $\frac{1}{3}$

2. $\frac{49}{36} = 1\frac{13}{36}$

3. $\frac{28}{5} = 5\frac{3}{5}$

4. $\frac{37}{4} = 9\frac{1}{4}$

5. $2\frac{17}{90}$

Page 59 Quick Test

1.

x	−2	−1	0	1	2	3
y	−11	−8	−5	−2	1	4

Page 61 Quick Test

1. a) Gradient = 3
   Intercept = 5
   b) Gradient = 6
   Intercept = −7
   c) Gradient = −3
   Intercept = 2

2.

x	−3	−2	−1	0	1	2	3
y	4	2	2	4	8	14	22

Page 62

1. Any quadrilateral-based pyramid, e.g. square-based pyramid **[1]**
2. a)

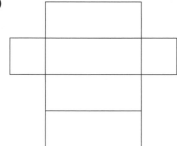

[1]

b)

[1]

c)

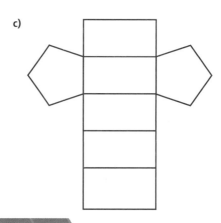

[1]

1. Surface area = $2(14 \times 2) + 2(6 \times 2) + 2(14 \times 6)$ **[1]** = 248 cm² **[1]**
   Volume = $14 \times 2 \times 6$ **[1]** = 168 cm³ **[1]**
2. Radius = 2.2 ÷ 2 = 1.1 **[1]**
   $\pi \times 1.1^2 \times 11$ **[1]** = 41.8 m³ **[1]**
3. $\sqrt[3]{512}$ **[1]** = 8 m = 800 cm **[1]**
4. Surface area = $2(15 \times 10 + 15 \times 20 + 10 \times 20) = 1300$ cm² **[1]**
   No, he has not got enough paper **[1]**

Page 63

1.

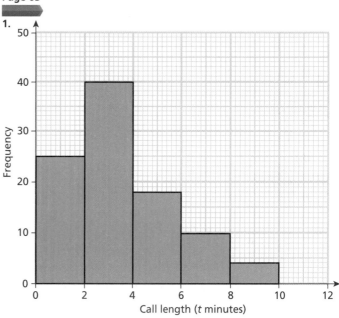

Axes labelled correctly **[1]**; all bars correct **[2]**; three bars correct **[1]**

1.

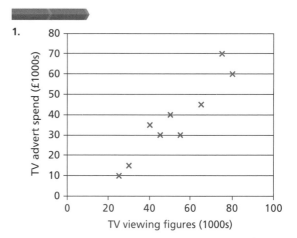

Correct axes labels **[1]**. Points plotted correctly **[1]**. It has a positive correlation **[1]**; as advertising spend increases, viewing figures also increase **[1]**

2. a) Example answers: 'a lot' is too vague – the quantity should be specific **[1]**; how much junk food in a period of time could be asked **[1]**

b) Example answers: Needs to cover all possible options, e.g. 0 and greater than 4 **[1]**; quantity of fruit could be more specific, e.g. portion size **[1]**

3.

x	−1	0	1	2	3	4
y	−7	−4	−1	2	5	8

All correct **[2]**; at least four correct **[1]**

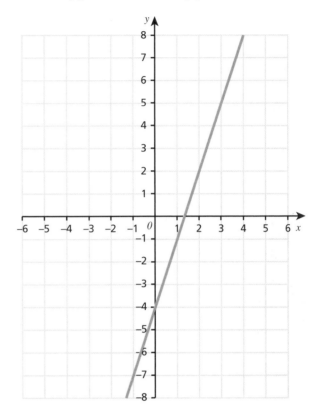

[1]

Pages 64–65 Practice Questions

Page 64

1.

$\frac{3}{5} = \frac{9}{15}$ **[1]**

$\frac{2}{3} = \frac{10}{15}$ **[1]**

So $\frac{2}{3}$ is greater than $\frac{3}{5}$ **[1]**

2. a) $\frac{21}{20} = 1\frac{1}{20}$ **[1]**

b) $\frac{41}{40} = 1\frac{1}{40}$ **[1]**

c) $\frac{29}{20} = 1\frac{9}{20}$ **[1]**

d) $\frac{5}{8}$ **[1]**

e) $\frac{3}{10}$ **[1]**

f) $\frac{19}{36}$ **[1]**

3. a) $\frac{2}{24} = \frac{1}{12}$ **[1]**

b) $\frac{40}{54} = \frac{20}{27}$ **[1]**

c) $\frac{3}{20}$ **[1]**

4. a) $\frac{3}{16}$ **[1]**

b) $\frac{9}{48} = \frac{3}{16}$ **[1]**

c) $\frac{21}{12} = \frac{7}{4} = 1\frac{3}{4}$ **[1]**

1. a) $\frac{35}{8} + \frac{11}{5} = \frac{263}{40}$ **[1]** $= 6\frac{23}{40}$ **[1]**

b) $\frac{18}{5} + \frac{21}{9} = \frac{267}{45}$ **[1]** $= 5\frac{14}{15}$ **[1]**

c) $\frac{29}{4} - \frac{30}{11} = \frac{199}{44}$ **[1]** $= 4\frac{23}{44}$ **[1]**

d) $\frac{11}{5} - \frac{10}{7} = \frac{77}{35} - \frac{50}{35}$ **[1]** $= \frac{27}{35}$ **[1]**

1.

x	−3	−2	−1	0	1	2	3
y	−2	−4	−4	−2	2	8	16

All correct **[3]**; at least five correct **[2]**; at least three correct **[1]**

Page 65

1. (1, 1) **[1]** and (4, 4) **[1]**

2.

[2]

[1] for each correct line.

Pages 66–73 Revise Questions

Page 67 Quick Test

1. Student's own drawings

2. a) 76°
b) 56°
c) 65°

Page 69 Quick Test

1. a) 55°
b) 112°
c) 126°

2. Equilateral triangle, square or hexagon

Page 71 Quick Test

1. Likely

2.

3. a) $\frac{1}{3}$ **b)** $\frac{2}{3}$

4. 0.15

Page 73 Quick Test

1. a) 0.4
b) 0.6
c) $0.2 \times 60 = 12$

Page 74

1. a) $\frac{3}{4} = \frac{6}{8} = \frac{12}{16} = \frac{60}{80}$ etc.   [1]

    b) $\frac{1}{4} = \frac{2}{8} = \frac{4}{16} = \frac{10}{40}$ etc.   [1]

    c) $\frac{3}{5} = \frac{6}{10} = \frac{12}{20}$ etc.   [1]

1. a) $\frac{4}{10} + \frac{1}{10} = \frac{5}{10} = \frac{1}{2}$ [1]

    b) $\frac{7}{12} + \frac{3}{12} = \frac{10}{12} = \frac{5}{6}$ [1]

    c) $\frac{5}{30} + \frac{6}{30} = \frac{11}{30}$ [1]

    d) $\frac{20}{70} + \frac{21}{70} = \frac{41}{70}$ [1]

    e) $\frac{8}{9} - \frac{3}{9} = \frac{5}{9}$ [1]

    f) $\frac{14}{22} - \frac{11}{22} = \frac{3}{22}$ [1]

    g) $\frac{27}{30} - \frac{20}{30} = \frac{7}{30}$ [1]

2. a) $\frac{4}{45}$ [1]

    b) $\frac{9}{70}$ [1]

    c) $\frac{10}{36} = \frac{5}{18}$ [1]

    d) $\frac{2}{9} \times \frac{4}{1} = \frac{8}{9}$ [1]

    e) $\frac{4}{5} \times \frac{11}{6} = \frac{44}{30} = \frac{22}{15} = 1\frac{7}{15}$ [1]

3. $\frac{2}{5} + \frac{1}{4} = \frac{8}{20} + \frac{5}{20} = \frac{13}{20}$

    $1 - \frac{13}{20}$ **[1]** $= \frac{7}{20}$ **[1]**

4. $\frac{4}{9} + \frac{1}{3} = \frac{7}{9}$ [1]

    $1 - \frac{7}{9} = \frac{2}{9}$ [1]

    Shared equally $= \frac{1}{9}$ chocolate   [1]

5. a) $\frac{77}{9}$ [1]

    b) $\frac{23}{7}$ [1]

    c) $\frac{14}{11}$ [1]

Page 75

1. and 2.

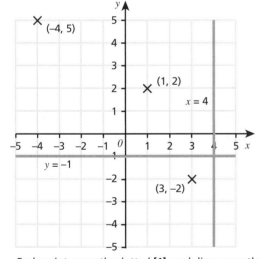

Each point correctly plotted **[1]**; each line correctly drawn **[1]**

1.

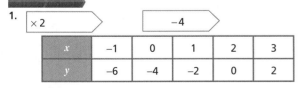

[1] for at least two correct values.

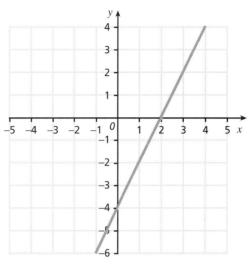

[1] for plotting at least two correct points.   **[2]**

2.

x	–3	–2	–1	0	1	2	3
y	–5	–5	–3	1	7	15	25

All correct **[3]**; At least five correct **[2]**; At least three correct **[1]**

Page 76

1. a) $180 - 146 = 34°$ [1]
    b) $76°$ [1]
    c) $90 - 56 = 34°$ [1]
2. $180 - 24 = 156°$ **[1]**, $\frac{156}{2} = 78°$ **[1]**
3. Heptagon   [1]

1. a) $x = 134°$ **[1]** as it is an alternate angle, $y = 180 - 134 = 46°$ **[1]**
    b) $x = 180 - 53 = 127°$ **[1]**; $y = 127°$ **[1]** as it is a corresponding angle
2. $180 - 150 = 30°$ **[1]** (exterior angle)
    $\frac{360}{30} = 12$ sides   [1]

Page 77

1. Impossible or 0   [1]
2. a) $\frac{3}{8}$ [1]
    b) $1 - \frac{1}{8}$ **[1]** $= \frac{7}{8}$ **[1]**

1. $1 - 0.65$ **[1]** $= 0.35$ **[1]**

    Probability of all outcomes minus the probability of it landing other way up.

2. a)

Number	Frequency	Estimated probability
1	5	$\frac{5}{50} = \frac{1}{10}$
2	8	$\frac{8}{50} = \frac{4}{25}$
3	7	$\frac{7}{50}$
4	7	$\frac{7}{50}$
5	8	$\frac{8}{50} = \frac{4}{25}$
6	15	$\frac{15}{50} = \frac{3}{10}$
	Total 50	1

All correct **[3]**; at least four rows correct **[2]**; at least two rows correct **[1]**

b) i) $\frac{3}{10}$ [1]

ii) $\frac{5}{50} + \frac{7}{50} + \frac{8}{50} = \frac{20}{50} = \frac{2}{5}$ [1]

iii) $\frac{15}{50} + \frac{8}{50} = \frac{23}{50}$ [1]

Pages 78–85 Revise Questions

Page 79 Quick Test
1. 0.35 and 35%   2. $\frac{9}{25}$
3. £28   4. $49

Page 81 Quick Test
1. £405
2. £92 000
3. 84%

Page 83 Quick Test
1. ÷ 6   2. 12
3. 5   4. $y = 5$
5. $x = -9$

Page 85 Quick Test
1. $x = 5$ 2. $x = 3$
3. $x = 2$ 4. 11
5. 2 kg

Pages 86-87 Review Questions

Page 86

1. a) $180 - 90 - 46 = 44°$ [1]
   b) $180 - 128 = 52°$ [1]
   c) Because it is an isosceles triangle, $180 - 38 = 142$, $\frac{142}{2} = 71°$ [1]

1. a) $x = 180 - 61 = 119°$ [1]
   $y = 119°$ (as it is corresponding) [1]
   b) $y = 180 - 114 = 66°$ [1]
   $x = 66°$ (as it is an alternate angle) [1]
2. $1260°$ [1]
3. $180 - 160 = 20°$ [1]
   $\frac{360}{20} = 18$, so it is an 18-sided shape. [1]

Page 87

1. From left: impossible, unlikely, even chance, likely, certain
   All correct [2]; three correct [1]

2. a) $\frac{3}{10}$ [1]
   b) $\frac{4}{10} + \frac{2}{10} = \frac{6}{10} = \frac{3}{5}$ [1]
   c) $1 - \frac{1}{10} = \frac{10}{10} - \frac{1}{10}$ [1] $= \frac{9}{10}$ [1]

1. a)

Sprinkles	Frequency	Probability
Chocolate	19	$\frac{19}{50} = 0.38$
Hundreds and thousands	14	$\frac{14}{50} = 0.28$
Strawberry	7	$\frac{7}{50} = 0.14$
Nuts	10	$\frac{10}{50} = 0.2$

All correct [2]; at least two rows correct [1]

b) $\frac{19}{50} + \frac{10}{50}$ [1] $= \frac{29}{50}$ or 0.58 [1]

2. $1 - 0.47 = 0.53$ [1]

3. a)

Sales destination	Probability of going to destination
London	0.26
Cardiff	0.15
Chester	0.2
Manchester	0.39

[1]

b) Cardiff (it has the lowest probability) [1]

Page 88

1.

Fraction	Decimal	Percentage	
$\frac{7}{10}$	0.7	70	[1]
$\frac{55}{100} = \frac{11}{20}$	0.55	55	[1]
$\frac{32}{100} = \frac{8}{25}$	0.32	32	[1]
$\frac{3}{100}$	0.03	3	[1]

2. a) 50% of £32 = 32 ÷ 2 = £16 [1]
   b) 10% of 80 cm = 80 ÷ 10 = 8 cm [1]
   c) 10% of 160 m = 160 ÷ 10 = 16 m
      5% = 16 ÷ 2 = 8 m [1]
      15% = 16 + 8 = 24 m [1]
   d) 50% = 52
      25% = £26 [1]
3. a) £17 [1]   b) £80 [1]   c) 15 m [1]

1. a) 15 ÷ 3 [1]   b) 210 ÷ 7 [1]   c) 6000 ÷ 5 [1]
      5 × 2 [1]   30 × 3 [1]   1200 × 4 [1]
      = £10 [1]   = £90 [1]   = £4800 [1]
2. 10% of £75 = 75 ÷ 10 = £7.50
   20% = £7.50 × 2 = £15 [1]
   Sale price = £75 − £15 [1]
   = £60 [1]
3. Karim gets $\frac{16}{20} = \frac{80}{100}$ [1]

   ×5

   = 80% [1]

   John gets $\frac{15}{20} = \frac{75}{100}$ [1]

   ×5

   = 75% [1]

Page 89

1. a) △ = 10 [1]
   b) △ = 5 [1]
   c) $n = 12$ [1]
   d) $y = 7$ [1]
2. a) $3n + 1 = 13$
      (−1) $3n = 12$ [1]
      (÷3) $n = 4$ [1]
   b) $2x - 5 = 3$
      (+5) $2x = 8$ [1]
      (÷2) $x = 4$ [1]
   c) $5y + 1 = 11$
      (−1) $5y = 10$ [1]
      (÷5) $y = 2$ [1]

1. a) $3x + 1 = x + 7$
      (−x) $2x + 1 = 7$
      (−1) $2x = 6$ [1]
      (÷2) $x = 3$ [1]
   b) $2(2x - 3) = x - 3$
               $4x - 6 = x - 3$
      (−x) $3x - 6 = -3$ [1]
      (+6) $3x = 3$
      (÷3) $x = 1$ [1]
   c) $6(x + 1) = 2(x + 13)$
               $6x + 6 = 2x + 26$
      (−2x) $4x + 6 = 26$ [1]
      (−6) $4x = 20$
      (÷4) $x = 5$ [1]

d) $\dfrac{3x + 5}{4} = 5$

   (×4) $3x + 5 = 20$ **[1]**

   (−5) $3x = 15$

   (÷3) $x = 5$ **[1]**

2. $3n + 2 = 11$ **[1]**

   (−2) $3n = 9$

   (÷3) $n = 3$ **[1]**

3. $n + 16 = 28$ **[1]**

   (−16) $n = 12$ **[1]**

Pages 90–97 Revise Questions

Page 91 Quick Test

1. 6

2. a) 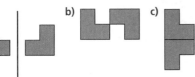   **b)**   **c)**

3. Rectangle $3\,\text{cm} \times 6\,\text{cm}$

Page 93 Quick Test

1. Any three shapes exactly the same size and shape.

2. D   **3.** 4–5 m

4. A and B   **5.** Any two similar shapes

Page 95 Quick Test

1. a) 6 : 8   **b)** 8 : 6

2. a) 1 : 3   **b)** 7 : 1   **c)** 1 : 4

Page 97 Quick Test

1. 10 : 25 : 5   **2.** Sara £200, John £160

3. a) 20   **b)** 28   **4.** £40

Pages 98–99 Review Questions

Page 98

1. a) $\frac{3}{20} = \frac{15}{100} = 15\%$ **[1]**

  b) $0.8 = \frac{8}{10}$ **[1]** $= \frac{4}{5}$ **[1]**

2. a) $23 \div 2 = £11.50$ **[1]**

  b) $45 \div 10 = 4.5\,\text{cm}$ **[1]**

  c) $180 \div 4 = 45\,\text{m}$ **[1]**

  d) $10\% = 70 \div 10 = 7$ **[1]**

     $5\% = 7 \div 2 = £3.50$ **[1]**

3. a) £14 **[1]**

  b) £240 **[1]**

  c) £680 **[1]**

1. a) $5 \div 5 \times 2$ **[1]**

    $= £2$ **[1]**

  b) $£5 - (£2.50 + £2)$ **[1]**

    $= £0.50$ or 50 pence **[1]**

2. 10% of £90 $= 90 \div 10 = £9$

    $5\% = £9 \div 2 = £4.50$ **[1]**

    $15\% = £9 + £4.50 = £13.50$ **[1]**

    Sale price $= £90 - £13.50$

           $= £76.50$ **[1]**

3. $150 \div 100 \times 6$

    $6\% = £9$ **[1]**

    After four years $£9 \times 4 = £36$ **[1]**

    Total in account $= £186$ **[1]**

Page 99

1. a) ⬤ $= 4$ **[1]**

  b) $n = 9$ **[1]**

  c) $p = 18$ **[1]**

  d) ⬤ $= 19$ **[1]**

2. a) $4n - 1 = 11$

   (+1) $4n = 12$ **[1]**

   (÷4) $n = 3$ **[1]**

b) $5x + 1 = 21$

   (−1) $5x = 20$ **[1]**

   (÷5) $x = 4$ **[1]**

c) $3a + 8 = 5$

   (−8) $3a = -3$ **[1]**

   (÷3) $a = -1$ **[1]**

1. a) $6x - 5 = 4x + 7$

   (−4x) $2x - 5 = 7$

   (+5) $2x = 12$ **[1]**

   (÷2) $x = 6$ **[1]**

b) $5(x + 2) = 2(x - 1)$

        $5x + 10 = 2x - 2$

   (−2x) $3x + 10 = -2$ **[1]**

   (−10) $3x = -12$

   (÷3) $x = -4$ **[1]**

c) $3x - 1 = 4 - 2x$

   (+2x) $5x - 1 = 4$ **[1]**

   (+1) $5x = 5$

   (÷5) $x = 1$ **[1]**

2. $\dfrac{20n + 150}{5} = 50$ **[1]**

   (×5) $20n + 150 = 250$

   (−150) $20n = 100$

   (÷20) $n = 5$

   The builders worked for 5 hours **[1]**

3. $56 - n = 29$ **[1]**

   $n = 56 - 29 = 27$

   27 chocolate bars were sold **[1]**

Pages 100–101 Practice Questions

Page 100

1.

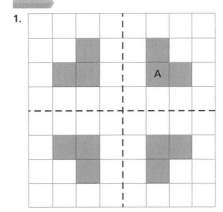

              [3]

[1] for each correct shape.

2. Order 2 **[1]**

1.

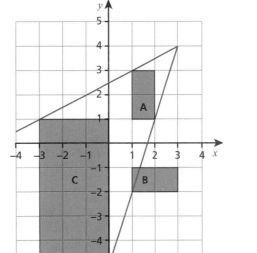

a) See diagram [2]
   [1] for any 90° rotation.
b) See diagram [2]
   [1] for any enlargement of shape A or B by scale factor 3
c) A and B [1]
2. 5 cm × 4 = 20 cm [1]
   7 cm × 4 = 28 cm [1]

Page 101

1. 3 : 6 **OR** 1 : 2 [1]
2. **a)** 1 : 3 [1]
   b) 6 : 1 [1]
   c) 20 cm : 100 cm [1]
      1 : 5 [1]
   d) 80 mins : 90 mins [1]
      8 : 9 [1]

1. 450 ÷ 9 = 50 [1]
   Ann: 4 × 50 = £200 [1]
   Ben: 5 × 50 = £250 [1]
2. 3 parts = £27
   1 part = 27 ÷ 3 = £9 [1]
   2 parts = £9 × 2 = £18 [1]
   Total sum of money = £27 + £18 = £45 [1]
3. Butter: 40 ÷ 6 × 12 = 80 g [1]
   Flour: 100 ÷ 6 × 12 = 200 g [1]
4. **a)** 2.5 × 50 000 = 125 000 cm [1]
      = 1250 m
      = 1.25 km [1]
   b) 1.4 × 50 000 = 70 000 cm [1]
      = 700 m
      = 0.7 km [1]

Pages 102–109 **Revise Questions**

Page 103 Quick Test
1. **a)** 40 km   **b)** 16 km
2. **a)** 19 miles   **b)** 25 miles
3. **a)** £3   **b)** £4
c)

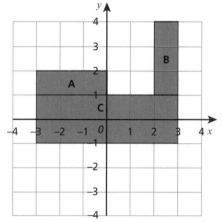

Page 105 Quick Test
1. 1.5 hours
2. 80 km/h
3. Kamala (21 km/h, John 20 km/h)
4. €240
5. 250 g

Page 107 Quick Test
1. **a)** 10.24   **b)** 244.9225
2. **a)** 70   **b)** 6.3

3. 5.46 cm
4. 7.94 cm

Page 109 Quick Test
1. **a)** 0.3420   **b)** 0.8660   **c)** 1
2. **a)** 56.5°   **b)** 55.2°   **c)** 88.2°

Pages 110–111 **Review Questions**

Page 110

1.

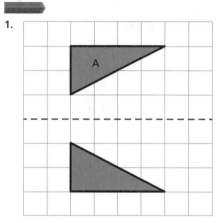

[2]

[1] for any reflection in a horizontal line.

1. **a)** Order 1 [1]   **b)** Order 4 [1]   **c)** Order 2 [1]
2. **a)** See diagram [2]
      [1] for any 90° rotation.
   b) See diagram [2]
      [1] for any enlargement of shape A or B by scale factor 2

c) A and B [1]
3. 3 cm : 6 m = 3 cm : 600 cm [1] = 1 : 200 [1]

Page 111

1. 16 : 14 [1]
   8 : 7 [1]
2. **a)** 5 : 1 [1]
   b) 2 : 3 [1]
   c) 25 pence : 200 pence [1]
      1 : 8 [1]
3. **a)** 200 [1]
   b) 10 [1]
   c) 20 [1]

1. **a)** **40** : 15 [1]
   b) 7 : **12** [1]
2. 4 parts = £120
   1 part = 120 ÷ 4 = £30 [1]
   Altogether there is £120 + £30 [1]
   = £150 [1]

3. $40 \div 8 = 5$ **[1]**
3 parts $= 5 \times 3 = 15$
5 parts $= 5 \times 5 = 25$
15 pens : 25 pens   **[1]**
4. 1 postcard $= £2.16 \div 18 = 12$ pence or £0.12   **[1]**
27 postcards cost $£0.12 \times 27 = £3.24$ **[1]**
5. $12 \div 60 \times 90$ **[1]** $= 18$ foot shadow **[1]**

Pages 112–113 **Practice Questions**

Page 112

1. 20×60 **[1]**
$= 1200$ mph   **[1]**
2. 3.50×1.19 **[1]**
$= €4.165$ **[1]**

1. France, with correct reasoning, e.g.   **[1]**
Converts $ and € to £
$1\,000\,000 \div 2.7 = £370\,370$ **[1]**
$780\,000 \div 1.54 = £506\,494$
OR
Converts $ to €
$1\,000\,000 \div 2.7 \times 1.54 = 570\,370$
OR
Converts € to $
$780\,000 \div 1.54 \times 2.7 = 1\,367\,532$
2. a) Speed = distance ÷ time
$= 350$ km $\div 1.1$ h   **[1]**
$= 318$ km/h   **[1]**
b) Accept 2115 – 2118 (9:15 – 9:18pm)   **[1]**

Page 113

1. a) 289   **[1]**
b) 12.25   **[1]**
c) 23   **[1]**
d) 6.4   **[1]**

1. a) $9^2 + AC^2 = 17^2$
$AC = \sqrt{289 - 81} = \sqrt{208}$ **[1]**
$AC = 14.4$ cm   **[1]**
b) $2^2 + 5^2 = BC^2$
$BC = \sqrt{29}$ **[1]**
$BC = 5.39$ m   **[1]**
2. a) $\sin 25° = \frac{P}{17}$
$P = \sin 25° \times 17$ **[1]**
$P = 7.18$ m   **[1]**
b) $\tan y° = \frac{32}{46}$
$y = \tan^{-1}(32 \div 46)$ **[1]**
$y = 34.8°$ **[1]**
3. $\cos x° = \frac{15}{22}$ **[1]**
$x = \cos^{-1}(15 \div 22)$ **[1]**
$x = 47°$ **[1]**

Pages 114–115 **Review Questions**

Page 114

1. a) $200 \div 1.75$ **[1]**
$= £114.29$ **[1]**
b) 200×1.75 **[1]**
$= US\$350$ **[1]**
2. a) Time $= 300 \div 60$ **[1]**
Time $= 5$ hours   **[1]**
b) Distance $= 60 \times 4$ **[1]**
$= 240$ miles   **[1]**

1. Density $= 2000 \div 5$ **[1]**
Density $= 400$ kg/m³   **[1]**

2.

Name	Journey description
Sanjay	This person walked slowly and then ran at a constant speed.
Dee	This person walked at a constant speed but turned back for a while before continuing.
Ann	This person walked at a constant speed without stopping or turning back.
Ben	This person walked at a constant speed but stopped for a while in the middle.

All correct **[2]**; two correct **[1]**

Page 115

1. a) 10.89   **[1]**
b) 14   **[1]**
2. a) $6^2 + 8^2 = y^2$ **b)** $23^2 + 15^2 = x^2$
$y = \sqrt{100}$ **[1]** $x = \sqrt{754}$ **[1]**
$y = 10$ cm **[1]** $x = 27.5$ m **[1]**

1. a) $\tan x° = \frac{15}{8}$ **b)** $6^2 + AB^2 = 9^2$
$x = \tan^{-1}(15 \div 8)$ **[1]** $AB = \sqrt{81 - 36} = \sqrt{45}$ **[1]**
$x = 61.9°$ **[1]** $AB = 6.7$ cm **[1]**
2. a) $8^2 + 2^2 = y^2$ **[1]** for using '2'
$y = \sqrt{68}$ **[1]**
$y = 8.2$ cm **[1]**
b) Perimeter $= 8.2 + 8.2 + 4 = 20.4$ cm (or 20.5 using unrounded values)   **[1]**

Pages 116–127 **Mixed Test-Style Questions**

Pages 116–121 No Calculator Allowed

1. a) Surface area $= 2(6 \times 4 + 6 \times 2 + 4 \times 2) = 88$ cm²   **[1]**
Volume $= 6 \times 4 \times 2 = 48$ cm³   **[1]**
b) Surface area $= 2(12 \times 7 + 12 \times 8 + 7 \times 8) = 472$ cm²   **[1]**
Volume $= 12 \times 7 \times 8 = 672$ cm³   **[1]**
2. a) $4\frac{1}{2} + 2\frac{1}{3} = \frac{9}{2} + \frac{7}{3} = 6\frac{5}{6}$ **[1]**
b) $5\frac{2}{3} + 8\frac{1}{4} = \frac{17}{3} + \frac{33}{4} = 13\frac{11}{12}$ **[1]**
c) $9\frac{1}{6} - 2\frac{3}{8} = \frac{55}{6} - \frac{19}{8} = 6\frac{19}{24}$ **[1]**
d) $12\frac{1}{2} - 14\frac{5}{6} = \frac{25}{2} - \frac{89}{6} = -2\frac{1}{3}$ **[1]**

3.

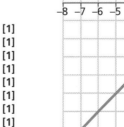

    [3]

4. a)

x	−2	−1	0	1	2	3
y	−9	−6	−3	0	3	6

[2] for all values of y correct; **[1]** for three or more correct

b)

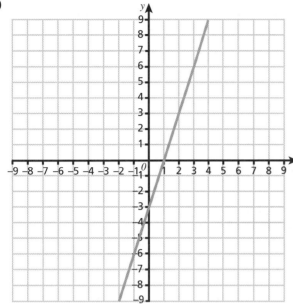

Correctly plotted line **[2]**
Straight line passing through one of the correct coordinates **[1]**
5. a) $x = 115°$ (corresponding and opposite) **[1]**
   $y = 55°$ (alternate) **[1]**
b) $y = 180 − 85 = 95°$ **[1]**, $x = 180 − 75 = 105°$ **[1]**
6. a) $4x + 4y$ **[1]**
b) $3g + 1$ **[1]**
7. a) $4x − 20$ **[1]**
b) $4x^2 + 16x$ **[1]**
8. a) $6(x − 2)$ **[1]**
b) $4x(x − 2)$ **[1]**
9. $z = 2$ **[3]**
   [2] if 9 is seen; **[1]** for a correct attempt to find the area of the trapezium with no more than one numerical error
10. 120 bars **[3]**
   [2] for $\frac{7200}{60}$ or $\frac{72}{0.6}$ seen; **[1]** for 7200 or 72 seen

> Remember to work in pounds or pence, not a mixture of the two.

11. a) 16 or 64
b) 15 or 45 or 51 or 54 or 57 or 75
c) 17 or 41 or 47 or 61 or 67 or 71
   [2] for all three correct; **[1]** for any two correct
12. a) $400 + 40 = 440$ mm **[1]**
b) $440 + 44 = 484$ mm **[1]**
13. a) Cone **[1]**
b) Tetrahedron **[1]**
c) Cylinder **[1]**
14.

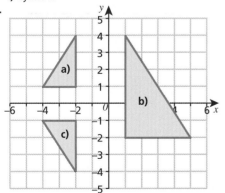

[3]

Pages 122–127 Calculator Allowed
1. a)

Type	Frequency	Probability	
Tug boat	12	$\frac{12}{40} = \frac{3}{10}$	**[1]**
Ferry boat	2	$\frac{2}{40} = \frac{1}{20}$	**[1]**
Sail boat	16	$\frac{16}{40} = \frac{2}{5}$	**[1]**
Speed boat	10	$\frac{10}{40} = \frac{1}{4}$	**[1]**

b) $\frac{2}{5} \times 75 = 30$ **[1]**
2. a) Surface area $= 326.7$ cm² **[2]**
   [1] for $8 \times \pi \times 9 + 2(4^2 \times \pi)$
   Volume $= 452.4$ cm³ **[2]**
   [1] for $4^2 \times \pi \times 9$
b) Surface area $= 298.5$ cm² **[2]**
   [1] for $10 \times \pi \times 4.5 + 2(5^2 \times \pi)$
   Volume $= 353.4$ cm³ **[2]**
   [1] for $5^2 \times \pi \times 4.5$
3. 3.43 cm² (to 2 d.p.) **[3]**
   [2] for 16 **and** $\pi(2^2)$ or 12.566 37 seen;
   [1] for 16 **or** $\pi(2^2)$ or 12.566 37 seen

> Find the area of the square and subtract the area of the circle. Remember the diameter of the circle is the same as the side length of the square, in this case 4 cm.

4. Carol's Cars and £6000 and £6050 seen **[3]**
   [2] for £6000 and £6050 seen but no conclusion; **[1]** for £5400 seen
5. a) 8.333 m/s **[2]**
   [1] for $100 ÷ 12$ seen
b) 30 km/h **[2]**
   [1] for 0.1 km $÷ 0.003 333$ h seen
6. 10.6 m **[2]**
   [1] for $\sin 45° \times 15$
7. Any answer from 82–84.5 km **[3]**
   [2] for approx. 20 km + 22 km + 41.2 km seen;
   [1] for two correct measurements in km
8. a) 0.08 **[1]**
b) $\frac{2}{25}$ **[1]**
9. a) 160 g of butter **[1]**, 400 g of flour **[1]**
b) 200 g of butter **[1]**, 500 g of flour **[1]**
10. a) 7 **[1]**
b) 16 **[1]**
11. a) 28 minutes **[1]**
b) Median is the 8th value when listed in order **[1]**
   Median = 28 minutes **[1]**
c) $49 − 10 = 39$ minutes **[1]**

Glossary

a

alternate angles   angles that lie on opposite sides of a transversal and are equal in size

angle   the space (usually measured in degrees) formed by two intersecting lines or surfaces

arc   a curve; part of the circumference of a circle

area   the space inside a 2D shape

arithmetic sequence   a sequence of numbers with a common difference

axis   a line that provides scale on a graph; often referred to as x-axis (horizontal) and y-axis (vertical)

b

bar chart   a chart that uses horizontal or vertical bars of equal width to represent statistics

biased   a statistical event where the outcomes are not equally likely

BIDMAS   the order in which operations should be carried out: Brackets; Indices; Division and Multiplication; Addition and Subtraction

binomial   an algebraic expression of the sum or difference of two terms

bisect   to cut exactly in two

brackets   symbols used to enclose a sum

c

centre of enlargement   the position from which a shape is enlarged

centre of rotation   the point about which a shape is rotated

certain   an outcome of an event which must happen, probability equals 1

circle   a round 2D shape

circumference   the perimeter of a circle

class interval   in grouped data, the width of the group (difference between the upper and lower limit of the group)

coefficient   the number in front of a variable in a term, e.g. in the term $3a$ the coefficient is 3

combined events   when two or more events take place

common difference   the number added or subtracted at each stage of an arithmetic sequence

common factor   a number that can be divided into two different numbers, without leaving a remainder

composite   a complex 2D or 3D shape made from several simpler shapes

conditional probability   probability that is affected by an earlier outcome

congruent   exactly the same

constant   a value that does not change

conversion graph   a graph used to convert one unit to another

coordinates   usually given as (x, y); the x-value is the position horizontally, the y-value the position vertically

correlation   the relationship between data, the 'pattern'; can be positive or negative

corresponding angles   angles that are in the same position on two parallel lines relative to the transversal and are equal in size

cos (cosine)   the ratio of the adjacent side to the hypotenuse in a right-angled triangle

cross-section   a cross-section of a solid is a slice cut through the solid at right-angles to its axis

cube number   a number that can be expressed as the product of three equal integers

cube root   the inverse (or opposite) of cubing, i.e. the number that is multiplied by itself twice to form the cube

cylinder   a 3D shape with a circular top and base of the same size

d

data   a collection of answers or values linked to a question or subject

decimal   a number that contains tenths, hundredths, etc

decimal places   the number of digits after the decimal point

decrease   to make smaller

degree   a unit of measure of an angle

denominator   the bottom number of a fraction

density   the mass of something per unit of volume

diameter   the distance across a circle, going through the centre

difference   subtraction

direct proportion   quantities are in direct proportion if their ratio stays the same as the quantities increase or decrease

distance   length

divide   to share

double   multiply by 2

e

edge   a line segment joining two vertices in a 2D or 3D shape

enlargement   a transformation in which a shape is made bigger or smaller

equally likely   having the same chance of happening

equation   a mathematical statement containing an equals sign

equivalent   the same as

estimate   a simplified calculation (not exact), often rounding to 1 significant figure

even chance   an equally likely chance of an event happening or not happening

event   a set of possible outcomes from a particular experiment

expand   remove brackets by multiplying

experimental probability   the ratio of the number of times an outcome happens to the total number of trials

exponent   another word for the 'power' or 'index'; see *index*

exponential graph   a graph of the form $y = k^x$

expression   a collection of algebraic terms

exterior angle   an angle outside a polygon formed between one side and the adjacent side extended

f

face   a side of a 3D shape

factor   a number that divides exactly into another number

factorise   take out the highest common factor and add brackets

fair   not biased

formula   a rule linking two or more variables

fraction   any part of a number or 'whole'

frequency table   a table that shows the number of times 'something' occurs

function machine   a flow diagram that shows the order in which operations should be carried out

g

geometric sequence   a sequence of numbers in which each term is the product of the previous term multiplied by a constant value

gradient   the measure of steepness of a line

grouped data   data which has been collected in or sorted into groups

h

highest common factor   the highest factor two or more numbers have in common

hypotenuse   the longest side of a right-angled triangle

hypothesis   a prediction of an experiment or outcome

i

impossible   an outcome of an event which cannot happen, probability equals 0

improper fraction   a fraction where the numerator is larger than the denominator

income tax   tax paid on money earned

increase   to make bigger

index   the power to which a number is raised; in 2^4 the base is 2 and the index is 4

infinite   a word that describes a sequence that continues forever

integer   whole number

intercept   the point at which a graph crosses an axis

interest   a monetary amount added on to savings or loans

interior angle   the measure of an angle inside a shape

interpret   to describe the trends shown in a statistical diagram or statistical measure; the way in which a representation of information is used or surmised

inverse   the opposite of

inverse proportion   a relationship where one value increases as another value decreases so that their product is always equal

l

like terms   terms with the same variables

likely   a word used to describe a probability which is between evens and certain on a probability scale

line of best fit   the straight line (usually on a scatter graph) that represents the closest possible line to each point; shows the trend of the relationship

linear   in one direction, straight

locus   the only set of points that satisfies certain conditions

lowest common multiple   the lowest multiple two or more numbers have in common

lowest terms   a fraction or ratio in which the parts have no common factors

m

mean   a measure of average; sum of all the values divided by the number of values

median   a measure of average; the middle value when data is ordered

mixed number   a number with a whole part and a fraction

mode   a measure of average; the most common

multiplier   a number that you are multiplying by

mutually exclusive   events or outcomes that cannot happen at the same time

n

negative   less than zero

net   a 2D representation of a 3D shape, i.e. a 3D shape 'unfolded'

*n*th term   see *position-to-term*

numerator   the top number of a fraction

o

ordinary number   a number not written in standard form

outcome   a possible result of a probability experiment

outlier   a statistical value which does not fit with the rest of the data

p

parallel   lines are parallel if they are always the same distance apart; this means they will never meet

parallelogram   a quadrilateral with two pairs of equal and opposite parallel sides and equal opposite angles

percentage   out of 100

perimeter   distance around the outside of a 2D shape

perpendicular   at 90° to

perpendicular bisector   the line which cuts another line in half and is at right angles to it

pi (π)   the ratio between the diameter of a circle and its circumference, approx. 3.142

pictogram   a frequency diagram in which a picture or symbol is used to represent a particular frequency

pie chart   a circular diagram divided into sectors to represent data, where the angle at the centre is proportional to the frequency

place value   indicates the value of the digit depending on its position in the number

position-to-term   a rule which describes how to find a term from its position in a sequence

positive   greater than zero

power   see *index*

prime   a number with exactly two factors, itself and 1

prime factor decomposition   the process of breaking a number down into a product of prime factors

prism   a 3D shape with uniform cross-section

probability   the likeliness of an outcome happening in a given event

probability scale   a scale to measure how likely something is to happen, running from 0 (impossible) to 1 (certain)

product   multiplication

protractor   a piece of equipment used to measure angles

Pythagoras' Theorem   in a right-angled triangle, the square on the hypotenuse is equal to the sum of the squares of the other two sides

q

quadratic   based on square numbers

quadratic equation   an equation where the highest power of x is x^2

quadrilateral   a four-sided 2D shape

quantity   an amount

r

radius   half the diameter; the measurement from the centre of a circle to the edge

random   each possible outcome is equally likely

range   the difference between the biggest and smallest number in a set of data

ratio   a comparison of two amounts

raw data   original data as collected

ray   a line connecting corresponding vertices

reciprocal   the inverse of any number except zero, e.g. the reciprocal of 2 is $\frac{1}{2}$ and the reciprocal of $\frac{3}{4}$ is $\frac{4}{3}$

reciprocal graph   a graph of the form $y = \frac{1}{x}$

reflection   a mirror image

regular polygon   a 2D shape that has equal-length sides and equal angles

rotation   a turn

rounding   a number can be rounded (approximated) by writing it to a given number of decimal places or significant figures

s

sample space a way in which the outcomes of an event are shown

scale the ratio between two or more quantities

scale factor the ratio by which a shape/number has been increased or decreased

scatter graph paired observations plotted on a 2D graph

sector a section of a circle enclosed between an arc and two radii (a pie piece)

sequence a set of numbers or shapes which follow a given rule or pattern

share to divide

significant figures the importance of digits in a number relative to their position; in 3456 the two most significant figures are 3 and 4

similar two shapes that have the same shape but not the same size

simplify make simpler, normally by cancelling a fraction or ratio or by collecting like terms

simultaneous equations equations that represent lines that intersect

sin (sine) the ratio of the opposite side to the hypotenuse in a right-angled triangle

solve work out the value of

speed how fast something is moving

square a regular four-sided polygon; to multiply by itself

square number a number made from multiplying an integer by itself

square root the opposite of squaring; a number when multiplied by itself gives the original number

standard form a way of writing a large or small number using powers of 10, e.g. $120\,000 = 1.2 \times 10^5$

subject a single variable isolated on one side of a formula

substitute to replace a letter in an expression with a number

sum addition or total

supplementary angles two angles that add up to $180°$

surface area the total area of all the faces of a 3D shape

survey a set of questions used to collect information or data

t

tally chart a simple way of recording and showing information by using vertical marks to count each occurrence; every fifth tally mark is a diagonal line through the previous four

tan (tangent) the ratio of the opposite side to the adjacent side in a right-angled triangle

term a number that forms part of a sequence; in expressions, terms are separated by + and − signs

terminating decimal a decimal number with a finite number of digits after the decimal point, e.g. 0.75, 0.125

term-to-term the rule which describes how to move between consecutive terms

tessellation a pattern made by repeating 2D shapes with no overlap or gap

transversal a line that crosses two parallel lines

trapezium a quadrilateral with only one pair of parallel sides

triangle a three-sided 2D shape

triangular numbers numbers that can be represented by a triangular pattern of dots, e.g. 10

u

units these define length, speed, time, volume, etc

unknown a number that is not known

unlikely a word used to describe a probability which is between evens and impossible on a probability scale

v

value added tax (VAT) tax paid when certain items are purchased

variable a quantity that can take on a range of values, usually denoted by a letter such as x or y

vertex the point where two or more edges meet on a 2D or 3D shape

vertical line graph a diagram that is similar to a bar chart but uses lines rather than bars

vertically opposite angles equal angles that face each other at a point

volume the capacity, or space, inside a 3D shape

Index

Graph Paper

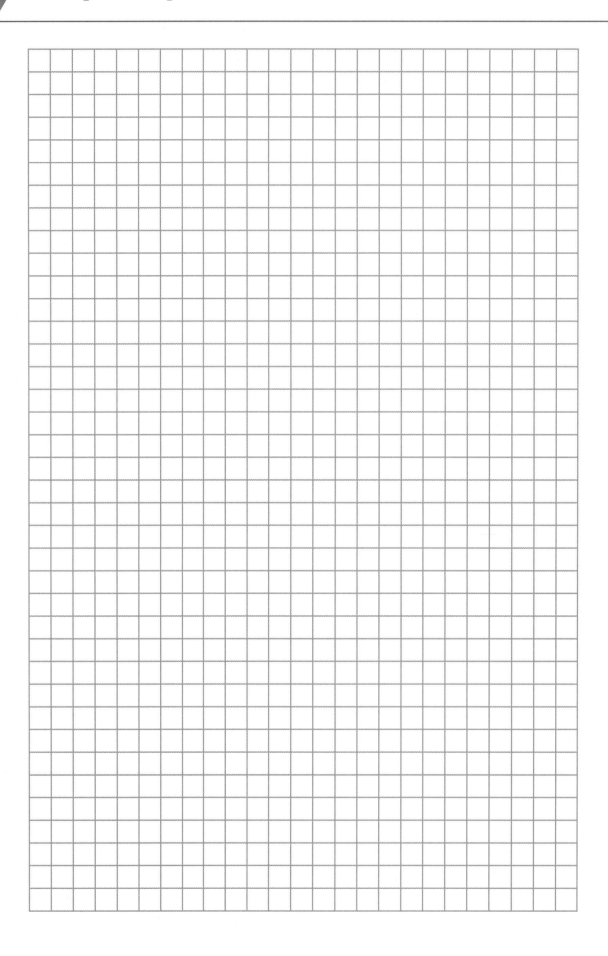

Collins

KS3
Maths
Foundation Level
Workbook

Trevor Senior

Contents

Rethink Revision

Have you ever taken part in a quiz and thought '*I know this*!', but no matter how hard you scrabbled around in your brain you just couldn't come up with the answer?

It's very frustrating when this happens, but in a fun situation it doesn't really matter. However, in tests and assessments, it is essential that you can recall the relevant information when you need to.

Most students think that revision is about making sure you **know** *stuff*, but it is also about being confident that you can **retain** that *stuff* over time and **recall** it when needed.

Revision that Really Works

Experts have found that there are two techniques that help with *all* of these things and consistently produce better results in tests and exams compared to other revision techniques.

Applying these techniques to your KS3 revision will ensure you get better results in tests and assessments and will have all the relevant knowledge at your fingertips when you start studying for your GCSEs.

It really isn't rocket science either – you simply need to:

- **test yourself** on each topic as many times as possible
- **leave a gap** between the test sessions.

It is most effective if you leave a good period of time between the test sessions, e.g. between a week and a month. The idea is that just as you start to forget the information, you force yourself to recall it again, keeping it fresh in your mind.

Three Essential Revision Tips

1 Use Your Time Wisely
- Allow yourself plenty of time
- Try to start revising six months before tests and assessments – it's more effective and less stressful
- Your revision time is precious so use it wisely – using the techniques described on this page will ensure you revise effectively and efficiently and get the best results
- Don't waste time re-reading the same information over and over again – it's time-consuming and not effective!

2 Make a Plan
- Identify all the topics you need to revise (this Complete Revision & Practice book will help you)
- Plan at least five sessions for each topic
- A one-hour session should be ample to test yourself on the key ideas for a topic
- Spread out the practice sessions for each topic – the optimum time to leave between each session is about one month but, if this isn't possible, just make the gaps as big as realistically possible.

3 Test Yourself
- Methods for testing yourself include: quizzes, practice questions, flashcards, past papers, explaining a topic to someone else, etc.
- This Complete Revision & Practice book gives you seven practice test opportunities per topic
- Don't worry if you get an answer wrong – provided you check what the right answer is, you are more likely to get the same or similar questions right in future!

Visit our website to download your free flashcards, for more information about the benefits of these revision techniques and for further guidance on how to plan ahead and make them work for you.

collins.co.uk/collinsks3revision

Number

1 Complete the boxes to make each calculation correct.

a) $14 + 8 - 2 - \boxed{} = 0$　　b) $74 - 24 - \boxed{} = -1$　　c) $(8 + 13) - 8 - \boxed{} = 0$　　[3]

2 Write these temperatures in order from coldest to warmest.

7°C　　　　　　−2°C　　　　　　−3°C　　　　　　4°C　　　　　　0°C　　　　　　−8°C

.... [2]

3 Write **true** or **false** for each statement.

a) $-5 > 2$ ....

b) $-6 < -8$ ....

c) $3 > -3$ ....

d) $-1 < -7$ ....

e) $-1 \neq -1$ .... [5]

4 Write each number in its correct place on the diagram. 5, 8, 16, 31, 64

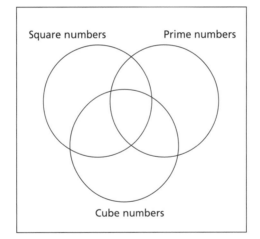

[5]

Total Marks .... / 15

(MR) **5** A two-digit cube number and a square number have a difference of 2

What are the two numbers?

.... and .... [2]

(PS) **6** Write down two numbers that are factors of 24 and are also factors of 27

.... and .... [2]

(MR) **7** a) Place ticks in the cells of the table to show if the number is a multiple of 2, 3, 4 or 5

	Multiple of 2	Multiple of 3	Multiple of 4	Multiple of 5
80				
81				
82				

[4]

b) Find a number that is a common multiple of 2, 3, 4 and 5

.... [2]

(MR) **8** Decide whether each statement is **always true**, **sometimes true** or **never true**.

a) Multiples of 3 are also multiples of 9 .... [1]

b) Multiples of 12 are also multiples of 6 .... [1]

c) Adding three consecutive even numbers will give a multiple of 4 .... [1]

(PS) **9** A red car and a blue car go around a toy track. They start together from the same point. The red car takes 8 seconds to complete a circuit. The blue car takes 10 seconds.

After how many seconds will they next pass the start point at the same time?

.... s [2]

10 The Venn diagram shows the prime factors of 84 and 108

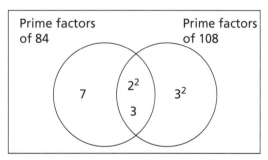

What is the highest common factor of 84 and 108?
Show how you worked out your answer.

.... [2]

Total Marks .... / 17

Sequences

1 This shape is made using 8 short sticks and 4 long sticks.

A sequence of patterns is made using short sticks and long sticks as shown.

  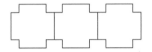 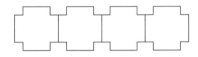

a) How many of the shapes are in pattern 10?

_____ [1]

b) How many short sticks are in pattern 5?

_____ [2]

c) How many long sticks are in pattern 6?

_____ [2]

d) How many long sticks are in the pattern that has 80 short sticks?

_____ [3]

2 For each sequence, write down the next two terms and the rule you used.

a) 22, 19, 16, 13, ..., ...

_____ [2]

b) 64, 32, 16, 8, ..., ...

_____ [2]

c) 2, 4, 6, 10, ..., ...

_____ [2]

d) 4, 5, 7, 10, ..., ...

_____ [2]

Total Marks _____ / 16

(MR) **3** Here are four sequences.

A 10, 20, 30, 40, ...   **B** 100, 95, 90, 85, ...

C 0.1, 0.01, 0.001, 0.0001, ...   **D** $\frac{1}{9}, \frac{1}{7}, \frac{1}{5}, \frac{1}{3}, ...$

a) Which of the sequences will contain the number 80?

.... [1]

b) Which of the sequences will contain negative numbers?

.... [1]

c) What position will 50 be in sequence B?

.... [2]

(MR) **4** Here are the nth terms of eight sequences.

Which of the sequences decrease? Circle your answers.

A $3n + 2$ **B** $3n - 2$ **C** $-3n + 2$ **D** $-3n - 2$

E $\frac{1}{3}n + 2$ **F** $\frac{1}{3}n - 2$ **G** $-\frac{1}{3}n + 2$ **H** $-\frac{1}{3}n - 2$

[2]

5 Here is a sequence.   5, 9, 13, 17, 21, ...

Jack thinks the 10th term of this sequence is 42

Explain why he is **not** correct.

....

[2]

6 Work out the nth term of the sequence   3, 9, 15, 21, ...

.... [2]

Total Marks .... / 10

Perimeter and Area

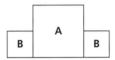

1 **a)** Work out the perimeter of square A and the perimeter of square B. 🖩

A   10 cm   B 5 cm

Perimeter of square A = _____ cm

Perimeter of square B = _____ cm   [2]

b) Work out the area of square A and the area of square B.

Area of square A = _____ cm²

Area of square B = _____ cm²   [2]

(PS) **c)** Two of square B are attached to square A, as shown.

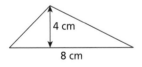

Work out the perimeter of the composite shape.

_____ cm   [2]

2 Work out the area of this triangle.

4 cm
8 cm

_____ cm²   [2]

(PS) **3** Here are a rectangle and a square. They have the same perimeter.

12 cm
8 cm

Work out the difference in the area of the two shapes.

_____ cm²   [3]

Total Marks _____ / 11

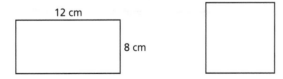

4 Here are five triangles.

Which of the triangles have the same area as triangle A?

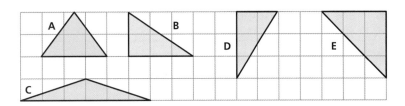

.... [2]

5 Two triangles X and Y are joined together to form a trapezium, as shown.

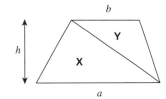

a) Write down a formula for the area of triangle X. .... [1]

b) Write down a formula for the area of triangle Y. .... [1]

c) Use your answers to parts a) and b) to show that the area of the trapezium is $\frac{1}{2}(a+b) \times h$

....

.... [1]

6 Use the fact that a circle with radius 5 cm has circumference 31.4 cm.

a) What is the circumference of a circle of radius 15 cm?

.... cm [2]

b) What is the perimeter of a semicircle of radius 5 cm?

.... cm [2]

7 a) Work out the area of a circle of diameter 10 cm.

.... cm^2 [2]

b) Work out the unshaded area.

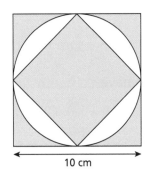

10 cm

.... cm^2 [3]

Total Marks .... / 14

Statistics and Data

1 Here are nine numbers.

6	3	8	4	13	8	4	4	13

a) Write down the mode. .... [1]

b) Work out the median. .... [1]

c) Work out the mean.

.... [2]

d) Work out the range. .... [1]

2 Data is collected about types of vehicle outside a school. The results are shown in a tally chart. Two more cars and two more lorries pass the school.

Add this information to the tally chart and then complete the frequency column.

Type of vehicle	Tally	Frequency
Car or van	JHT JHT JHT JHT	
Lorry or bus	JHT IIII	
Bicycle	II	
Other	JHT III	

[3]

PS **3** A teacher is giving out handfuls of coloured pencils to some students.
These are the number of pencils that each student gets.

Student	A	B	C	D	E	F
Number of pencils	8	12	11	15	9	11

The teacher tells the students to make sure that they each have the same number of pencils.

How many should each person have?

.... [2]

Total Marks .... / 10

4 Each week for four weeks, Anji spends £40 on fuel. On the fifth week, she spends £60.

What is the mean amount she spent on fuel over the five weeks?

£ .... [3]

(MR) 5 The table shows the number of tickets sold for a pantomime.

	Stall	Circle	Balcony	Total
Adult	240		192	
Child	160	147		
Total		300		1000

a) Complete the table. [2]

b) How many more adult tickets were sold than child tickets?

.... [2]

c) Adult tickets cost £25 and child tickets cost £20

How much money was taken from selling tickets?

£ .... [2]

6 The table shows information about the number of text messages sent by 30 students in one day.

Number of texts	Frequency
0–9	16
10–19	6
20–29	7
30 or more	1

a) What is the modal class? .... [1]

b) One of those students sent over 200 texts.

Why is the mean **not** suitable to use as an average for this data?

.... [1]

Total Marks .... / 11

Decimals

1 Choose the correct answer for each part. 10^3 10^2 30 10^0

a) $10 \times 10 \times 10$ .... [1]   **b)** $10 + 10 + 10$ .... [1]

c) $1 \times 10 \times 10$ .... [1]   **d)** 1 .... [1]

2 Complete the table so that the values in each column are equivalent.

Power of 10	10^{-1}		10^{-2}	
Fraction		$\dfrac{1}{1000}$		
Decimal				0.0001

[4]

3 For each diagram, fill in the missing number so that the bottom row has the same total value as the top number.

a) 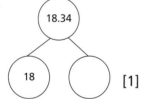 [1]   **b)** 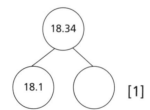 [1]   **c)** 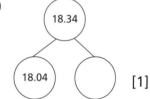 [1]

(MR) **4** Fill in the missing digits in these calculations.

a)
```
    4   0   2 . □ □ □
  - 1   2   9 . 3 2 5
  _____
    □ □ □ . 5 1 1
```

b)
```
    8   2   4 . □ 7 5
  - 6   9   □ . 4 2 6
  _____
    □   3   2 . 7 □ □
```

5 Write the missing number to make each calculation correct.

a) $0.2 \div \boxed{} = 0.002$ **b)** $4.32 \times \boxed{} = 0.432$ **c)** $999 + 99 + 9 = \boxed{}$ [3]

6 Work out:

a) 27×0.1 .... [1]   **b)** $27 \times \dfrac{1}{10}$ .... [1]

c) 27×10^{-1} .... [1]   **d)** 8.3×0.01 .... [1]

e) $8.3 \times \dfrac{1}{100}$ .... [1]   **f)** 8.3×10^{-2} .... [1]

Total Marks .... / 26

(MR) **7** Show that this statement is correct. $5 \times (13 \times 2.6 - 13 \times 0.6) = 10^2 + 10 \times 3$

....

.... [3]

8 **a)** Round 83 to the nearest 10 .... [1]

b) Round 83 to 1 significant figure. .... [1]

c) Round 274 to the nearest 100 .... [1]

d) Round 274 to 1 significant figure. .... [1]

e) I am thinking of a number. The number rounded to 1 significant figure is 6000

What is the smallest number I could be thinking of?

.... [1]

9 Complete the table by rounding each number to the given number of significant figures.

	1 s.f.	2 s.f.	3 s.f.	4 s.f.
1407				
2999				

[4]

10 A square has side 38.9 cm

a) Estimate the perimeter.

.... cm [2]

b) Estimate the area.

.... cm^2 [2]

11 Given that $6.4 \times 1.5 = 9.6$, work out:

a) 12.8×1.5 .... [2]

b) 6.4×2.5 .... [2]

Total Marks .... / 20

Algebra

1 **a)** Simplify $a + a + a + a$

[1]

b) Simplify $b \times b \times b$

[1]

c) Simplify $3x + 4y - x + 2y$

[1]

d) Simplify $2x^2 + 8y - x^2 - 11y$

[2]

2 Work out the missing coefficients.

a) $9x + \boxed{}\, y - \boxed{}\, x + 7y = 3x + 12y$ [2]

b) $7w + 5z - w - \boxed{}\, z = \boxed{}\, w - 2z$ [2]

3 Work out the value of the expression $4a + 6b - 2c$ when $a = 3$, $b = 2$ and $c = 1$

[2]

(MR) **4** The formula for working out the exterior angle, E, of a shape is $E = \dfrac{360}{n}$, where E is in degrees and n is the number of sides.

a) Work out the exterior angle of a regular pentagon.

° [1]

b) Show that the regular shape with exterior angle 36° is a decagon.

[1]

Total Marks _____ / 13

5 d is the number of days and h is the number of hours.

| $d = 24h$ |
| $d = 7h$ |
| $h = 7d$ |
| $h = 24d$ |

Which of the following shows the relationship between the number of hours and the number of days? Explain your answer.

[1]

6 a) Complete this algebraic multiplication table.

×	3a	4
5a		
4	12a	

[3]

b) Use your answer to part a) to multiply out and simplify $5a(3a + 4b) + 4(3a + 4b)$

[1]

7 Use the distributive law to write equivalent expressions for each of the following.

a) $3(x + 8)$ [1] **b)** $30(x + 8)$ [1]

c) $3a(x + 8)$ [1] **d)** $3ab(x + 8)$ [1]

8 Multiply out and simplify:

a) $(a + 3)(a + 4)$ [2]

b) $(b - 2)(b + 6)$ [2]

c) $(c + 5)^2$ [2]

9 Andy wants to work out the product $(3x + 4)(2x + 5y + 6)$

This is his answer.

What has Andy done wrong? Complete his answer.

×	3x	4
2x	$6x^2$	8x
5y	15xy	20y
6	18x	24

[1]

Total Marks / 16

3D Shapes: Volume and Surface Area

1   Complete the table.

Shape	Name	Number of faces	Number of edges	Number of vertices
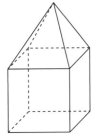	Cuboid			
	Tetrahedron			
	Cylinder			
	Triangular prism			

[4]

2   This solid is made from a cube and a square-based pyramid.

Write down the number of:

a) faces ....   **b)** edges ....   **c)** vertices ....   [3]

3   Draw a net of a square-based pyramid.   [1]

4   Work out the surface area and the volume of this cuboid.

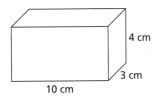

4 cm

3 cm

10 cm

Surface area = ... cm^2

Volume = ... cm^3 [4]

Total Marks .... / 12

(PS) **5** The diagram shows a prism.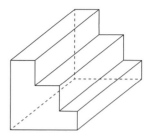

a) How many of the faces are rectangular? _____ [1]

b) How many faces are there altogether? _____ [1]

6 a) Draw labelled sketches of all the faces of this prism.

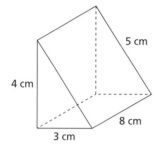

5 cm

4 cm

8 cm

3 cm

[4]

b) Use your sketches to work out the total surface area of the prism.

_____ cm² [4]

(PS) **7** A cylinder and a cuboid are shown.

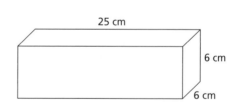

8 cm

25 cm

6 cm

6 cm

6 cm

a) Work out the volume of the cylinder. Give your answer to 1 decimal place.

_____ cm³ [2]

b) Work out the total surface area of the cylinder. Give your answer to 1 decimal place.

_____ cm² [2]

c) How many of the cylinders will fit inside the cuboid?

_____ [1]

Total Marks _____ / 15

Interpreting Data

1 The table shows the favourite food that 30 students eat for breakfast.

Food	Cereal	Toast	Cooked	Fruit	Other/None
Number of students	10	7	3	6	4

a) Complete this table to show the angle sizes that represent the information in a pie chart.

Number of students	30	1	3	4	6	7	10
Angle size	360°						

[3]

b) Draw the pie chart to represent the information.

Favourite Food

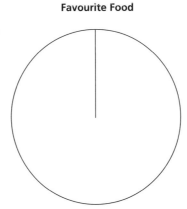

[2]

2 The table shows the time (to the nearest minute) taken by 30 people to get to work.

Time (nearest minute)	1–10	11–20	21–30	31–40	41–50
Frequency	5	7	9	8	1

Complete the chart to show this information.

[2]

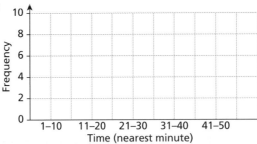

3 Describe the correlation for each scatter graph.

a)

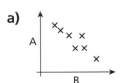

b)

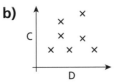

c)

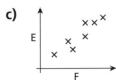

_____ correlation   _____ correlation   _____ correlation   [3]

Total Marks _____ / 10

(MR) **4** The bar chart shows the number of students in four year groups in a school.

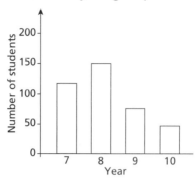

a) Which year group is the mode? .... [1]

b) If Year 11 was included, Year 11 would be the mode.

What can you say about the number of students in Year 11?

.... [1]

5 The pie chart shows the number of pieces of homework that 24 students are given one day.

Work out the mean number of pieces of homework per student.

Number of Homeworks

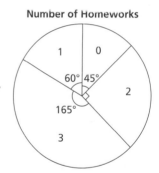

.... [4]

(MR) **6** The table shows the age of some randomly selected animals on two farms.

Farm A	8	7	8	9	7	6	5	9	9	7	8	5
Farm B	4	10	8	8	8	9	11	10	8	8	10	8

Ted says that farm B has a bigger range so the ages are less consistent. Is he correct?

Show your working, commenting on any outliers.

....

.... [2]

Total Marks .... / 8

Fractions

1 Write down a fraction that is greater than $\frac{1}{3}$ but less than $\frac{1}{2}$

_____ [1]

2 Which of these fractions are equivalent to $\frac{3}{8}$?

$\frac{6}{11}$ $\qquad$ $\frac{6}{16}$ $\qquad$ $\frac{12}{32}$ $\qquad$ $\frac{15}{42}$ $\qquad$ $\frac{24}{64}$

_____ [2]

3 **a)** Shade $\frac{5}{8}$ of the top grid and $\frac{3}{5}$ of the bottom grid.

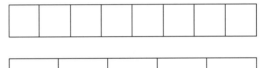

[2]

b) Which is greater, $\frac{5}{8}$ or $\frac{3}{5}$?

_____ [1]

4 Work out each of the following. Simplify where possible.

a) $\frac{1}{4} + \frac{1}{4}$ _____ [1]

b) $\frac{3}{5} + \frac{1}{5}$ _____ [1]

c) $\frac{5}{6} - \frac{1}{6}$ _____ [1]

d) $\frac{3}{8} + \frac{1}{4}$

_____ [2]

e) $\frac{7}{9} - \frac{2}{3}$

_____ [2]

f) $\frac{9}{10} + \frac{1}{5} - \frac{2}{3}$

_____ [2]

Total Marks _____ / 15

5 The diagram shows how $\frac{1}{2} \times \frac{1}{5} = \frac{1}{10}$

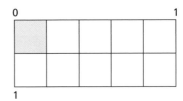

Use this grid to work out $\frac{1}{2} \times \frac{3}{5}$

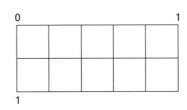

.... [2]

6 Insert $<$, $>$ or $=$ into each box to make the statements correct.

a) $5 \times \frac{1}{3}$ ☐ $\frac{5}{3}$ [1]   b) $\frac{1}{3} \times 4$ ☐ $1\frac{2}{3}$ [1]

c) $\frac{1}{5}$ ☐ $\frac{1}{2} \times \frac{1}{3}$ [1]   d) $\frac{1}{3} \times \frac{1}{2}$ ☐ $\frac{2}{5}$ [1]

7 Work out:

a) $\frac{1}{3} \times \frac{1}{4}$ .... [1]   b) $5 \times \frac{1}{3} \times \frac{1}{4}$ .... [1]

c) $\frac{2}{3} \times \frac{3}{4}$ .... [1]   d) $\frac{2}{3} \times \frac{3}{4} \times \frac{4}{5}$ .... [1]

 8 Use the fact that $\frac{7}{9} \times 234 = 182$ to work out:

a) $\frac{14}{9} \times 234$ .... [2]

b) $\frac{7}{3} \times 234$

.... [2]

c) $\frac{7}{18} \times 234$

.... [2]

9 Matt is flying to Malaga in Spain. The flight will take $2\frac{1}{2}$ hours. He checks the time and sees he is 45 minutes into the flight.

How long is it to the end of the flight? Give your answer as a mixed number in hours.

.... hours   [2]

Total Marks .... / 18

Coordinates and Graphs

1. Complete the table.

Equation	Gradient	Coordinates of y-intercept
$y = 4x - 3$	4	$(0, -3)$
$y = -2x + 6$		
$y = -2$		
$y = -8x$		
$y = -9 + 0.4x$		
$-0.5x - 0.6 = y$		

[5]

(MR) 2. Match each line to its equation.

$y = 2$ $\qquad$ $x + 2 = 0$ $\qquad$ $y = 2x$ $\qquad$ $y = x + 2$ $\qquad$ $y = -x + 2$ $\qquad$ $y = -2x$

a) A has equation .... [1]

b) B has equation .... [1]

c) C has equation .... [1]

d) D has equation .... [1]

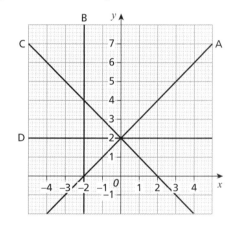

(MR) 3. This graph shows the line $y = \frac{1}{2}x - 1$

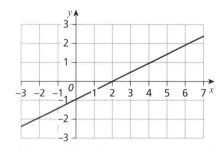

a) Write down the coordinates of a point where $y = \frac{1}{2}x - 1$ $\qquad$ (............ , ...) [1]

b) Write down the coordinates of a point where $y < \frac{1}{2}x - 1$ $\qquad$ (............ , ...) [1]

c) Write down the coordinates of a point where $y > \frac{1}{2}x - 1$ $\qquad$ (............ , ...) [1]

Total Marks .... / 12

PS **4** Here are the coordinates of three points. (4, 11) (5, 13) (7, 17)

a) Work out the relationship between the x and y values for this set.

.... [2]

b) Does a straight line pass through all three points?

.... [1]

MR **5** Here are some coordinates. (−5, −5) (−1, 3) (4, 13)

Circle the correct equation of the line passing through all three coordinates.

A $y = x$ **B** $y = 3x + 6$ **C** $y = 3x + 1$ **D** $y = 2x + 5$ [1]

MR **6** The points (−4, 2), (−2, 4) and (8, 14) lie on a straight line.

Ethan thinks the equation of the line passing through the coordinates is $x − 6 = y$

Explain why Ethan is wrong.

....

.... [2]

7 a) Complete the table of values for $y = x^2 − 2$

x	−3	−2	−1	0	1	2	3
y	7				−1	2	

[2]

b) On the grid, draw the graph of
$y = x^2 − 2$ from $x = −3$ to $x = 3$

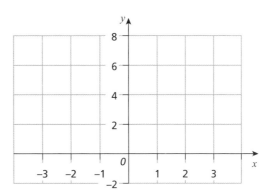

[2]

Total Marks .... / 10

Angles

① For each diagram, work out the size of the lettered angle(s). Give a reason for each answer.

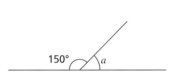
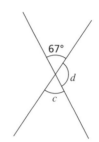
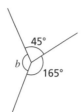

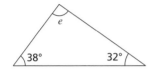
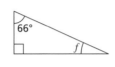
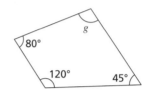

$a =$ .... °   Reason: ....

$b =$ .... °   Reason: ....

$c =$ .... °   Reason: ....

$d =$ .... °   Reason: ....

$e =$ .... °   Reason: ....

$f =$ .... °   Reason: ....

$g =$ .... °   Reason: .... [7]

② Look at the diagram.

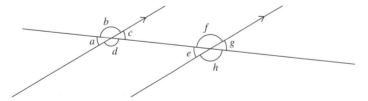

a) Write down all the angles that are equal in size to a. .... [1]

b) Write down all the angles that are equal in size to b. .... [1]

c) Write down a pair of alternate angles. .... and .... [1]

d) Write down a pair of corresponding angles. .... and .... [1]

Total Marks .... / 11

3 $a = 35°$. Work out the size of angles b and c. Give a reason for each answer.

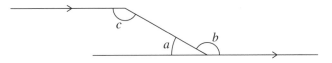

$b =$ .... °   Reason: ....

$c =$ .... °   Reason: .... [2]

4 Work out the size of all the lettered angles.

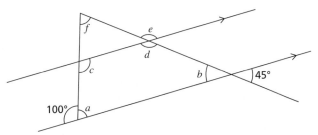

$a =$ .... ° $b =$ .... ° $c =$ .... °

$d =$ .... ° $e =$ .... ° $f =$ .... °   [6]

(MR) **5** Circles intersect as shown. A quadrilateral is formed using the centres and the points of intersection as vertices. In each case give the special name of the quadrilateral.

a) These circles have the same radii.   **b)**   **c)**

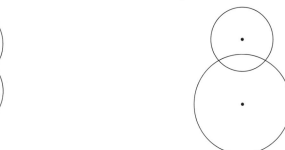 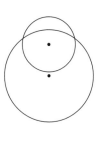

.... [3]

6 By splitting the regular octagon into triangles, work out the sum of the interior angles.

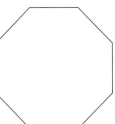

.... °   [2]

Total Marks .... / 13

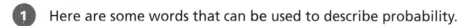
Probability

1   Here are some words that can be used to describe probability.

| Evens | Certain | Likely | Unlikely | Impossible |

Match the words to the probability scale.

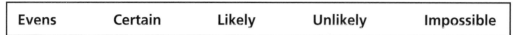

[4]

2   Ten cards numbered 1 to 10 are put into a bag. A card is then chosen at random.

a) What is the probability that the card has an odd number on it?

[1]

b) What is the probability that the card has a prime number on it?

[1]

c) What is the probability that the card does **not** have a multiple of 3 on it?

[1]

(MR) **3**   A fair dice is rolled five times. It lands: 2   3   6   6   6

The dice is thrown again. Is it more likely to land on 6 than any other number? Give a reason for your answer.

[1]

(PS) **4**   Four cards numbered as shown are turned face down.

Two cards are chosen at random and the numbers on them are added to give a score.

What is the most likely score?

[2]

Total Marks .... / 10

5 A fair six-sided spinner is shown. The arrow is spun.

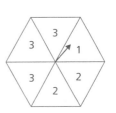

a) What is the probability that the arrow lands on 1? [1]

b) What is the probability that the arrow lands on 2? [1]

c) What is the probability that the arrow lands on 3? [1]

d) The spinner is now spun 50 times.

How many times do you expect the spinner to land on 3? [1]

6 Here is a Venn diagram.

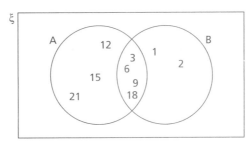

a) Write down the numbers that are in set A. [1]

b) Write down the numbers that are in set B′. [1]

c) Write down the numbers that are in set A ∪ B. [1]

d) Describe the numbers that are in set B.

[1]

e) A number is chosen at random from the Venn diagram.

What is the probability that it is in set A ∩ B? [1]

7 Two fair coins are thrown. They land on either heads (H) or tails (T).

a) List all the possible outcomes. The first one is done for you.

HH [1]

b) What is the probability of two heads? [1]

Total Marks _____ / 11

Fractions, Decimals and Percentages

Video Solution   Question 8

1 Complete the table of equivalent fractions, decimals and percentages.

Fraction	Decimal	Percentage
	0.7	
$\frac{2}{5}$		
		75%

[3]

2 Insert $<$ or $>$ in each box to make the statements correct.

a) 0.65 ☐ $\frac{2}{3}$ b) $\frac{4}{5}$ ☐ $\frac{2}{3}$ c) $\frac{4}{5}$ ☐ 0.65   [3]

3 Put the following in order, starting with the smallest: 0.1 $\frac{1}{6}$ $-\frac{1}{3}$ -0.3

_____   [2]

(MR) **4** Here are two number lines.

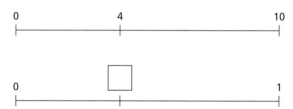

a) What is the missing value written as a decimal?   _____ [1]

b) What is the missing value written as a fraction?   _____ [1]

5 Work out $\frac{2}{3}$ of 60

_____ [2]

6 Work out 30% of 40 kg.

_____ kg   [2]

Total Marks _____ / 14

(FS) **7** **a)** Increase £40 by 30% £ .... [2]

b) Increase £30 by 40% £ .... [2]

c) Decrease £50 by 20% £ .... [2]

d) Decrease £60 by 70% £ .... [2]

(FS) **8** **a)** Match the **four** statements to the correct calculation. [4]

Increase £40 by 15%	Increase £15 by 40%	Decrease £40 by 15%	Decrease £15 by 40%

£15 × 1.4	£40 × 1.5	£40 × 1.15	£15 × 0.6	£40 × 0.6	£40 × 0.85

b) For the **two** calculations that have not been used, write a matching statement.

....

.... [2]

9 What is 15 out of 25 as a percentage?

.... % [2]

(FS) **10** A savings account pays 4% simple interest per year.

a) Jon pays £500 into the account.

How much interest will he be paid after 1 year?

£ .... [2]

b) How much will be in the account after 3 years?

£ .... [2]

Total Marks .... / 20

Equations

1 Solve the equations.

a) $a + 7 = 11$

b) $b - 9 = 15$

c) $4c = 24$

$a = $ ....

$b = $ ....

$c = $ ....

d) $\dfrac{d}{6} = 6$

e) $8e - 3 = 13$

f) $5f + 1 = 21$

$d = $ ....

$e = $ ....

$f = $ .... [8]

(PS) **2** Here are two boxes containing shapes. Each box contains the same total mass.

Any shapes that are identical have the same mass.

a) Are the hearts heavier or lighter than the triangles? Explain your answer.

.... [1]

b) The mass of one circle is 4 grams and the mass of one triangle is 6 grams.

Work out the mass of one heart.

.... g [3]

(MR) **3** **a)** Theo is thinking of a number. He multiplies his number by 5 and then subtracts 2
His answer is 28

What number is Theo thinking of?

.... [2]

b) Emily is thinking of a number. She subtracts 2 and then multiplies by 5
She says, "If I was thinking of the same number as Theo, I will get the same answer."

Is Emily correct? Show how you know.

....

.... [2]

Total Marks .... / 16

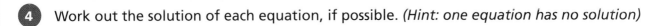

4 Work out the solution of each equation, if possible. *(Hint: one equation has no solution)*

a) $2x + 1 = 7$ ....

b) $3x + 4 = 2x + 4$ ....

c) $4x + 1 = 4x + 2$ ....

d) $5x = 7x + 4$ .... [7]

(MR) **5** Here is a set of equations.

$5x + 1 = 21$ $6x + 1 = x + 21$ $7x + 1 = 2x + 21$ $8x + 1 = 3x + 21$

a) What can you say about the solutions of these equations?

.... [1]

b) Write down the next equation in the sequence and its solution.

Equation: .... Solution: $x =$ .... [2]

6 Solve the equations.

a) $3x + 5 = 21 - x$ b) $\dfrac{4y + 7}{3} = 9$ c) $2(3z - 1) = 4(z + 2)$

$x =$ .... $y =$ .... $z =$ .... [9]

(PS) **7** The diagram shows an isosceles triangle.

The perimeter is 22 cm.

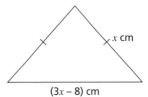
x cm
$(3x - 8)$ cm

Work out the value of x.

$x =$ .... [3]

Total Marks .... / 22

Symmetry and Enlargement

(MR) **1** Square ABCD is rotated to the image position shown.

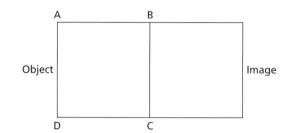

Describe fully **two** possible rotations that transform the object to the image.

_____

_____ [2]

2 Reflect each triangle in the dashed mirror line.

a)

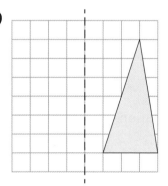

[1]

b)

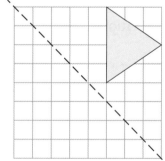

[1]

c)

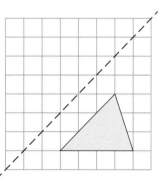

[2]

d)

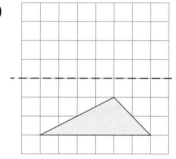

[2]

3 State the order of rotational symmetry of each shape.
Which shape does **not** have line symmetry?

_____ _____ _____ _____ _____ _____

Shape that does not have line symmetry: _____ [3]

Total Marks _____ / 11

(PS) **4** Triangles X, Y and Z are mathematically similar.

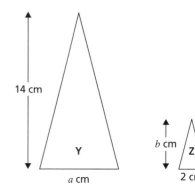

14 cm

7 cm

X

4 cm

Y

a cm

b cm

Z

2 cm

a) Work out the value of *a*. $a =$ .... [2]

b) Work out the value of *b*. $b =$ .... [2]

c) Write the ratio of the perimeters of X to Y to Z.

.... : .... : .... [1]

(MR) **5** Here are two mathematically similar parallelograms.

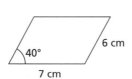

6 cm

40°

7 cm

x

9 cm

w

Decide whether each statement is **true** or **false**. If a statement is false, then correct it.

a) The scale factor of enlargement is +3

.... [1]

b) $w = 7 \times 1.5 = 10.5\,\text{cm}$

.... [1]

c) $x = 40° \times 1.5 = 60°$

.... [1]

6 A car is 4.5 metres long. A model of the car is built using a scale of 1 : 18

Work out the length of the model, in centimetres.

...........................cm [2]

Total Marks .... / 10

Ratio and Proportion

1 Write down the ratio of grey triangles to white triangles in its simplest form.

.... : .... [1]

(MR) **2** For these ratio tables, write down the multipliers and fill in the missing value.

Example:

2	6
8	

Horizontal multiplier is × 3; vertical multiplier is × 4

Missing number is 8 × 3 or 6 × 4 = 24

a)

4	8
20	

[2]

b)

6	9
15	

[2]

c)

8	10
6	

[2]

3 For each ratio table, write down the vertical multiplier and fill in the missing value.

a) Vertical multiplier: ....

b) Vertical multiplier: ....

c) Vertical multiplier: ....

£	%
30	100
	120

[2]

kg	%
10	100
	75

[2]

km	%
40	100
	125

[2]

(MR) **4** Each 4 by 6 grid is made of squares. Each square is to be coloured in red, blue or green.

a) Colour or label this grid so that the ratio
red : blue : green = 1 : 2 : 5

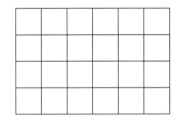

[2]

b) Is it possible to colour every square of this
grid if the ratio red : blue : green = 4 : 3 : 3 ?

Explain your answer.

....

.... [1]

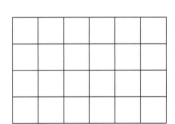

Total Marks .... / 16

5 150 people visit a park. The ratio of adults to children is 3 : 2

a) What fraction of the people are children?

.... [1]

b) How many adults are there?

.... [2]

6 The diameter of Mars to the diameter of the Moon is approximately in the ratio 2 : 1

If the diameter of the Moon is 3400 km, what is the diameter of Mars?

.... km [2]

(MR) **7** Matt and Tahir go for a walk. For every 4 steps Matt takes, Tahir takes 6 steps.

a) If Matt takes 240 steps, how many would Tahir take?

.... [2]

b) Natasha says, "If Tahir walks 960 steps, Matt will walk 640 steps."

Is she correct? Show your working.

.... [2]

(PS) **8** A pencil case contains red pens and blue pens in the ratio 1 : 4
There are 6 more blue pens than red pens.

How many blue pens are in the pencil case?

.... [2]

9 In a flapjack recipe that serves 4 people, there are 250 grams of oats.

How many grams of oats are needed for a recipe that serves 6 people?

.... g [2]

Total Marks .... / 13

Real-Life Graphs and Rates

FS **1** The exchange rate for pounds to euros is £1 = €1.20

a) Complete the table.

Pounds (£)	0	10	20	30	40	50
Euros (€)	0	12				

[2]

b) Draw the conversion graph.

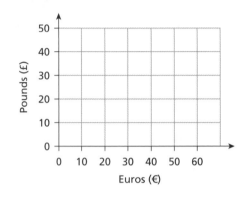

[2]

c) Use the graph to estimate the number of euros (€) that is equivalent to £35

€ .... [2]

d) Work out the number of pounds (£) that is equivalent to €900

£ .... [2]

FS **2** Terrier Taxis charge £4 for the first mile and then £1.50 per mile, as shown on the graph.

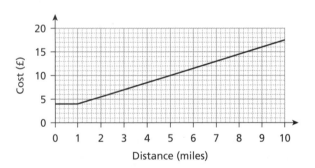

a) How much does a 10-mile journey cost? £ .... [1]

b) Call-a-Cab charges £2 for the first mile and then £2 per mile. Show this on the grid. [2]

c) Richard wants to book a taxi for a 7-mile journey.

Which company is cheaper?

.... [1]

Total Marks .... / 12

3 The distance–time graph shows Tariq's journey one afternoon.

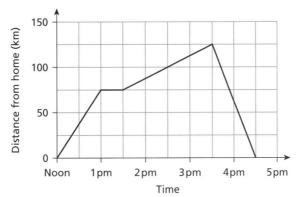

a) How long did Tariq stop for a break? .... minutes [1]

b) How far did he travel altogether?

.... km [2]

c) At what time did Tariq arrive home? .... [1]

d) Work out his average speed after his rest break and before his change in direction.

.... km/h [2]

PS **4** A car travels 190 miles at an average speed of 60 mph.

Did the journey take less than 3 hours? Show your working.

.... [2]

5 A piece of metal has a density of 7.5 g/cm^3. The mass of the metal is 60 grams.

Work out the volume.

.... cm^3 [2]

Total Marks .... / 10

Right-Angled Triangles

1 Use Pythagoras' Theorem to work out the unknown lengths.
Give your answers to 1 decimal place if they are not exact.

a)
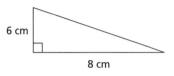
6 cm

8 cm

.... cm   [2]

b)
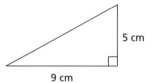
5 cm

9 cm

.... cm   [2]

c)
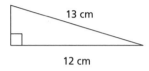
13 cm

12 cm

.... cm   [2]

d)
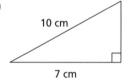
10 cm

7 cm

.... cm   [2]

2 Decide whether each triangle is right-angled. Show your working.

a)
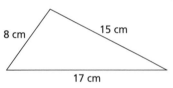
8 cm   15 cm

17 cm

.... [2]

b)
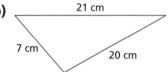
21 cm

7 cm   20 cm

.... [2]

c)
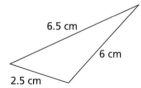
6.5 cm

6 cm

2.5 cm

.... [2]

3 Work out the perimeter of this triangle.

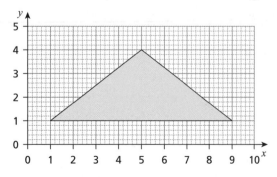

.... units   [4]

Total Marks .... / 18

4 Use your calculator to work out the value of the following.
Round your answers to 2 decimal places where necessary.

a) $\sin 72°$ .... [1]   **b)** $\cos 18°$ .... [1]

c) $9 \tan 20°$ .... [1]   **d)** $\dfrac{12}{\tan 50°}$ .... [1]

e) $\cos^{-1} 0.5$ .... [1]   **f)** $\tan^{-1} 0.466$ .... [1]

5 Work out the length of each lettered side. You are given that $\sin 40° = 0.6428$

a)

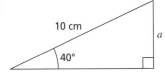

$a =$ .... cm   [2]

b)

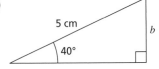

$b =$ .... cm   [2]

c)

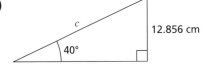

$c =$ .... cm   [2]

6 In each part, work out the size of angle x. Give your answers to 1 decimal place.

a)

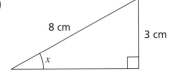

b)

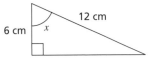

c)

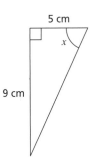

$x =$ .... ° [2] $x =$ .... ° [2] $x =$ .... ° [2]

Total Marks .... / 18

Mixed Test-Style Questions

No Calculator Allowed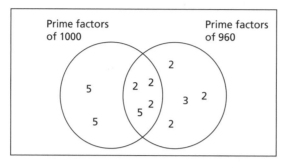

1 A Venn diagram is shown.

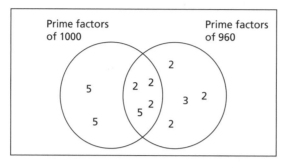

a) Use the Venn diagram to write 1000 as a product of prime factors.

2 marks

b) Use the Venn diagram to work out the highest common factor (HCF) of 1000 and 960

2 marks

2 **a)** Use the fact that $14 \times 243 = 3402$ to work out the answer to 15×243

2 marks

b) Work out $87 \times 24 + 87 \times 6$

2 marks

3 $a - b = -3$, where a and b are integers.

Write **true** or **false** for each statement.

a) $a > b$

1 mark

b) $a + 3 = b$

1 mark

c) b must be negative

1 mark

4 State whether each of these is **correct** or **incorrect**.

a) $12 \times 59 = 10 \times 50 + 2 \times 9$

....

1 mark

b) $12 \times 59 = 12 \times 60 - 12 \times 1$

....
1 mark

c) $12 \times 59 = 10 \times 59 + 2 \times 59$

....
1 mark

d) $12 \times 59 = 12 \times 50 + 12 \times 9$

....
1 mark

5 Jack is charged a booking fee of £4 for buying some tickets online. Each ticket costs £20.

Which calculation gives the total cost, in pounds? Circle your answer.

A Number of tickets $\times 24$ **B** Number of tickets $\times 80$

C Number of tickets $\times 4 + 20$ **D** Number of tickets $\times 20 + 4$

1 mark

6 State whether each of these is **correct** or **incorrect**.

a) $6025 \div 25 = (6025 \div 5) \div 5$

....

1 mark

b) $6025 \div 25 = (6025 \div 20) \div 5$

....

1 mark

c) $6025 \div 25 = (6025 \div 100) \times 4$

....
1 mark

7 These points lie on a straight line: (0, 5)   (1, 4)   (5, 0)

a) Work out the relationship between the x and y values for this set.

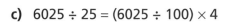
....
1 mark

b) Work out the coordinates of another point on the same line.

(............, ...)
1 mark

8 Which has the greater answer? $\frac{7}{8} \times \frac{8}{9}$ or $\frac{8}{9} \times \frac{9}{10}$

Show how you know.

2 marks

9 **a)** On this double number line, 10 and 8 are perfectly aligned.

Write down another pair of numbers that would be aligned.

1 mark

b) On this double number line, 16 and 4 are perfectly aligned.

What number is aligned with 18?

1 mark

10 Work out 38×27

2 marks

11 **a)** Simplify $4a + 8b + 6a - 7b$

2 marks

b) Expand and simplify $4(c + 8d) + 5(c - 7d)$

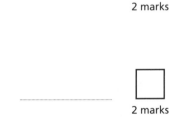

2 marks

12 Shapes A, B, C and D are shown on the grid.

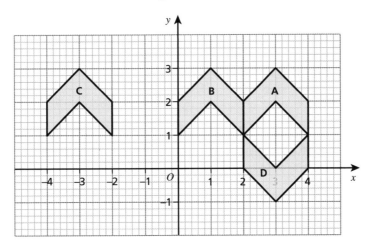

Describe fully the reflections that take:

a) shape A to shape B

b) shape A to shape C

c) shape A to shape D

13 Here is a sequence of patterns with answers.

Pattern 1 $2 + 4 = 6$

Pattern 2 $2 + 4 + 4 = 10$

Pattern 3 $2 + 4 + 4 + 4 = 14$

a) Write down the next row of the sequence.

b) Work out the answer to pattern 25

c) Which pattern has the answer 198?

Mixed Test-Style Questions

14 Here is a number machine.

a) Work out the output when the input is 7

1 mark

b) Work out the input when the output is –5

1 mark

15 The diagram shows a triangle with a line drawn parallel to the base.

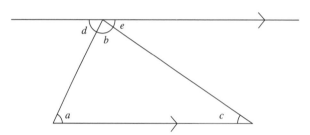

Fill in the spaces to make each statement correct.

$d =$ .... (alternate angles)

$e =$ .... (alternate angles)

$d + b + e = 180°$ (..)

so $a + b +$ .... $= 180°$

Therefore, the angles of a triangle add up to ....

5 marks

16 Construct a triangle with sides 3 cm, 6 cm and 8 cm.

2 marks

17 $(x + a)(x + b) = x^2 + ax + bx + 24$, where a and b are positive whole numbers.

a) Write down a pair of possible values for a and b.

$a =$ ...., $b =$ ....

1 mark

b) Write down a different pair of possible values for a and b.

$a =$ ...., $b =$ ....

1 mark

18 Given that $8.3 \times 2.5 = 20.75$, work out:

a) 83×25

....

2 marks

b) 8.3×12.5

....

2 marks

19 This graph shows the lines $y = \frac{1}{2}x - 1$ and $y = -x + 2$

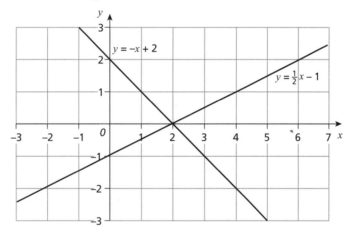

a) Write down the coordinates of a point on the grid where $y > \frac{1}{2}x - 1$ and $y < -x + 2$

(........................., ...)

1 mark

b) Write down the coordinates of a point on the grid where $y > \frac{1}{2}x - 1$ and $y = -x + 2$

(........................., ...)

1 mark

c) Write down the coordinates of the point on the grid where $y = \frac{1}{2}x - 1$ and $y = -x + 2$

(........................., ...)

1 mark

Mixed Test-Style Questions

20 Estimate the answer to:

a) 18.1×0.49

2 marks

b) $\dfrac{19.4 - 7.2}{2.1}$

2 marks

c) $\dfrac{9.6 + 4.31}{0.52}$

2 marks

d) $(28.2)^2$

2 marks

21 Given that $9x - 8 = 6x + 10$, state whether each equation is **true** or **false**.

a) $9x = 6x + 18$

1 mark

b) $8x = 5x + 18$

1 mark

c) $3x = 18$

1 mark

d) $x = 6$

1 mark

22 Circle the improper fraction that is equivalent to $4\frac{3}{5}$

$$\frac{43}{5} \qquad \frac{20}{5} \qquad \frac{23}{5} \qquad \frac{17}{5} \qquad \frac{37}{5}$$

1 mark

23 The shape on the grid is translated so that point A (1, 1) moves to point B (4, 3)

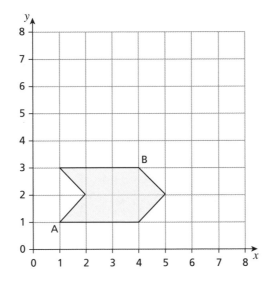

What are the coordinates of the point that B moves to?   (.............., ...)

2 marks

24 A bag of mathematical shapes contains 8 triangles, 6 circles, 4 squares and 2 rectangles.

A shape is chosen at random from the bag.

a) What is the probability that the shape is a square?
Give your answer as a fraction in its simplest form.

....

2 marks

b) What is the probability that the shape does **not** have four sides?
Give your answer as a fraction in its simplest form.

....

2 marks

c) One of each shape is now taken from the bag. Another shape is chosen at random.

Azuma says, "Because we have taken out the same number of each shape, the probabilities of choosing each particular shape is unchanged."

Is she correct? Give a reason for your answer.

....
....

1 mark

25 These are the marks for 10 students in a test.

8	7	9	2	8	4	8	10	3	6

a) Write down the mode.

_____ ☐

1 mark

b) Work out the median.

_____ ☐

2 marks

c) Work out the mean.

_____ ☐

2 marks

d) Work out the range.

_____ ☐

1 mark

26 The chart shows the attendance at two concerts.

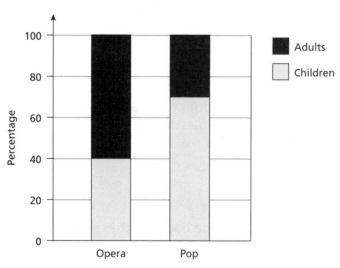

Compare the proportions of adults attending each concert.

_____

_____ ☐

1 mark

Calculator Allowed

1 Show the prime factors of 45 and 72 in the Venn diagram.

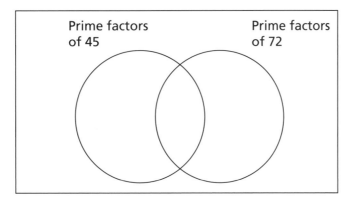

Prime factors of 45		Prime factors of 72	

3 marks

2 Which triangles are a rotation of triangle A?

A   B   C   D   E

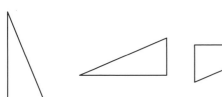

2 marks

3 Work out the size of angle x.

85°

x 45°

$x =$ _____ °

2 marks

4 The table shows the times (in seconds) for two people to complete the same task five times.

Amina	12	15	14	11	16
Bob	13	13	14	12	15

a) Give a reason why Amina could be faster.

_____

1 mark

b) Give a reason why Bob could be faster.

_____

1 mark

5 Lexi writes $(a + 9)^2 = a^2 + 81$

Ashima writes $(a + 9)^2 = a^2 + 9a + 9a + 81$

Rosie writes $(a + 9)^2 = a^2 + 18a + 81$

Who is correct?

....

2 marks

6 Here is a sequence.   20, 17, 14, 11, …
Jo thinks the nth term of the sequence is $3n + 17$

Is she correct? Give a reason for your answer.

....

2 marks

7 The diagram shows an arrowhead.

a) Draw the image of the arrowhead after rotating the object through 90°
anticlockwise about A.

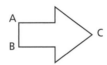

2 marks

b) Draw the image of the arrowhead after rotating the object through 90°
anticlockwise about C.

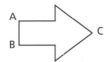

2 marks

c) Write down **one** feature that the images have in common.

....

1 mark

8 To estimate the adult height for a child:

Add together the height of the mother and father. For a boy add 13 cm and for a girl subtract 13 cm. Then divide the answer by 2.

a) Estimate the adult height of a boy whose parents' heights are 1.74 m and 1.52 m

.... m

2 marks

b) Estimate the adult height of a girl whose parents' heights are 1.82 m and 1.79 m

.... m

2 marks

9 Here are some coordinates.   (–6, –3) (–1, 2) (4, 7)

Asha thinks the equation of the line passing through these coordinates is $y = x - 3$

Explain why Asha is **not** correct.

....

1 mark

10 Decrease £280 by 60%

£.................................

2 marks

11 Solve the equations.

a) $8x - 7 = 65$

$x = $ ....

2 marks

b) $\frac{1}{2}x = 12$

$x = $ ....

1 mark

12 Work out the length x. Give your answer to 1 decimal place.

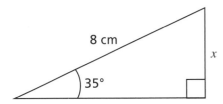

.... cm

2 marks

Mixed Test-Style Questions

13 A wooden plank has length 2 metres, width 15 cm and height 4 cm. It has a mass of 7.2 kg.

Work out the density. Give your answer in grams/cm³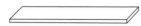

$$\text{density} = \frac{\text{mass}}{\text{volume}}$$

_____ g/cm³ ☐ 4 marks

14 The graph shows the exchange rate between the British pound (£) and the Hong Kong dollar (HK$).

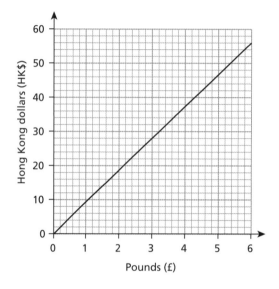

a) How many Hong Kong dollars would you get for £3? HK$ _____ ☐ 1 mark

b) How many Hong Kong dollars would you get for £900?

HK$ _____ ☐ 2 marks

c) What is the exchange rate from pounds to Hong Kong dollars?
Give your answer to 2 decimal places.

£1 = HK$ _____ ☐ 2 marks

15 The diagram shows a rectangle.

Work out the length of the diagonal.

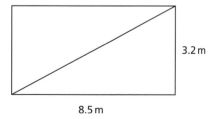

3.2 m

8.5 m

_____ m ☐ 3 marks

16 **a)** Factorise fully $4x^2 + 6x$

2 marks

b) Factorise $x^2 - 5x - 66$

2 marks

17 Three fair coins are thrown.

a) List all the possible outcomes of landing on heads (H) or tails (T). The first one is done for you.

HHH

2 marks

b) What is the probability of landing HHH?

1 mark

18 **a)** Complete the table of values for the graph of $y = 2x - 3$

x	−2	−1	0	1	2
y					

2 marks

b) Plot the graph of $y = 2x - 3$ for $x = -2$ to $x = 2$

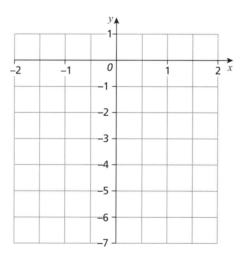

2 marks

c) Write down the coordinates of the y-intercept.

(............, ...)

1 mark

d) What is the gradient of the graph of $y = 2x - 3$?

1 mark

19 This shape is made from a rectangle and a semicircle.

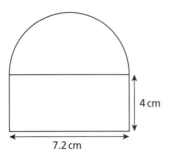

4 cm

7.2 cm

a) Work out the perimeter. Give your answer to 1 decimal place.

.... cm ☐

3 marks

b) Work out the area. Give your answer to 1 decimal place.

.... cm² ☐

3 marks

20 Work out the volume of this triangular prism.

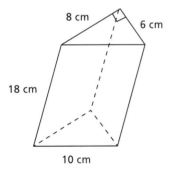

8 cm

6 cm

18 cm

10 cm

.... cm³ ☐

2 marks

21 A jet aircraft has length 36 metres. A model of the jet is made using a scale of 1 : 20

Work out the length of the model in centimetres.

.... cm ☐

2 marks

22 A shop sells carpet in 4 m wide rolls or 3 m wide rolls.

Pieces are only cut and sold in full widths.

The carpet costs £25.99 per square metre.

a) What is the cost of buying a piece 5 m long from the 4 m roll?

£ .... ☐

2 marks

b) What is the cheaper way to buy a carpet to fit an area measuring 3.7 m by 2.9 m?

.... ☐

2 marks

23 What is 340 g as a percentage of 2 kg?

.... % ☐

2 marks

24 Use your calculator to work out $\dfrac{848 \times 42}{96}$

Give your answer to 2 significant figures.

.... ☐

2 marks

25 Match each description to the correct graph.

| Exponential | | Direct proportion | | Inverse proportion |

☐

3 marks

26 The table shows the favourite sports of 40 students.

Sport	Football	Cricket	Tennis	Netball	Other / None
Number of students	17	8	3	5	7

a) Complete this table to show the angle sizes that represent the information in a pie chart.

Number of students	40	1	3	5	7	8	17
Angle size	360°						

3 marks

b) Draw the pie chart to represent the information.

Favourite Sport

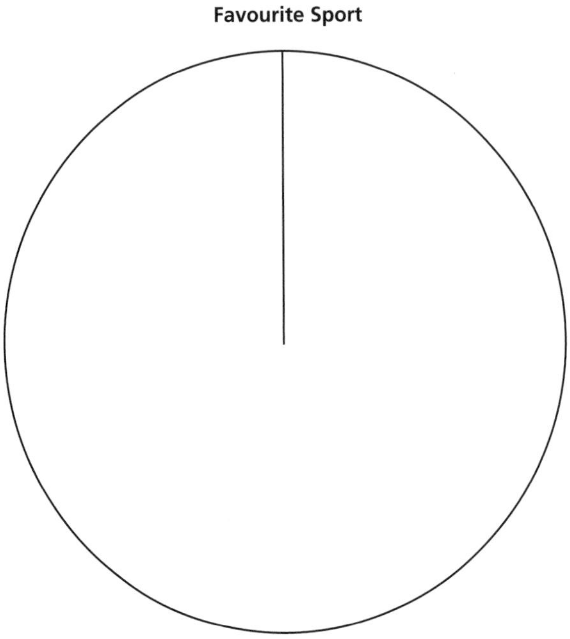

3 marks

Total Marks _____ / 82

Workbook Answers

Pages 148–149

Number

1. a) 20   [1]
    b) 51   [1]
    c) 13   [1]
2. −8°C, −3°C, −2°C, 0°C, 4°C, 7°C   [2]
    [1 mark if one error]
3. a) False   [1]
    b) False   [1]
    c) True   [1]
    d) False   [1]
    e) False   [1]
4.

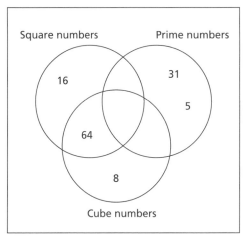

    [5]
    [1 mark for each number in the correct position]

5. 27 and 25   [2]
    [1 mark for listing at least three cube numbers (1, 8, 27, …) or at least three square numbers (1, 4, 9, …)]
6. 1 and 3   [2]
    [1 mark for one correct answer]
7. a)

	2	3	4	5
80	✓		✓	✓
81		✓		
82	✓			

     [4]

    [1 mark for each correct column]

b) 60 or any multiples of 60, e.g. 120, 180, …   [2]
[1 mark for $3 \times 4 \times 5$ or $2 \times 3 \times 4 \times 5$ or for listing multiples of 3, 4 and 5]

8. a) Sometimes true   [1]
    b) Always true   [1]
    c) Sometimes true   [1]
9. 40 s   [2]
    [1 mark for listing multiples of 8 and multiples of 10]
10. $2^2 \times 3$ [1]
    $= 12$ [1]

Pages 150–151

Sequences

1. a) 10   [1]
    b) 8×5 [1]
        $= 40$ [1]
    c) $4 + 5 \times 3$ or $4 + 15$ [1]
        $= 19$ [1]
    d) 80 short sticks are in pattern 10   [1]
        So long sticks in pattern $10 = 4 + 9 \times 3$ or $4 + 27$ [1]
        $= 31$ [1]
2. a) 10, 7   [1]
        Subtract 3 or − 3   [1]
    b) 4, 2   [1]
        Divide by 2 or ÷ 2   [1]
    c) 16, 26   [1]
        Add the two previous terms to get the next term (Fibonacci type)   [1]
    d) 14, 19   [1]
        +1, +2, +3, +4, +5, … or add one more each time   [1]

3. a) A and B   [1]
    b) B and D   [1]
    c) nth term is $105 - 5n$ or $105 - 5n = 50$ or $55 = 5n$ [1]
        $n = 11$ or 11th term   [1]

4. C, D, G and H [2]
    [1 mark for two correct and no incorrect answers]

5. nth term is $4n + 1$ or 10th term is $5 + 9 \times 4$ or $5 + 36$ [1]
    $= 41$ (Jack doubled the 5th term) so not correct [1]

6. The term-to-term rule is $+6$ [1]
    The difference between $6n$ and the output in each case is -3, so the nth term is $6n - 3$ [1]

Pages 152–153
Perimeter and Area

1. a) Perimeter of square A $= 4 \times 10 = 40\,\text{cm}$ [1]
       Perimeter of square B $= 4 \times 5 = 20\,\text{cm}$ [1]
    b) Area of square A $= 10 \times 10 = 100\,\text{cm}^2$ [1]
       Area of square B $= 5 \times 5 = 25\,\text{cm}^2$ [1]
    c) Perimeter =
       $10 + 5 + 5 + 5 + 5 + 10 + 5 + 5 + 5 + 5$ [1]
       $= 60\,\text{cm}$ [1]

2. Area of triangle $= \frac{1}{2} \times 8 \times 4$ [1]
    $= 16\,\text{cm}^2$ [1]

3. Perimeter of rectangle is
    $12 + 8 + 12 + 8 = 40\,\text{cm}$
    Length of one side of square is
    $40 \div 4 = 10\,\text{cm}$ [1]
    Area of rectangle is $12 \times 8 = 96\,\text{cm}^2$ and
    Area of square is $10 \times 10 = 100\,\text{cm}^2$ [1]
    Difference in area is $100 - 96 = 4\,\text{cm}^2$ [1]

4. Area of triangle A $= \frac{1}{2} \times 3 \times 2 =$
    3 square units
    Area of triangle B $= \frac{1}{2} \times 3 \times 2 =$
    3 square units
    Area of triangle C $= \frac{1}{2} \times 6 \times 1 =$
    3 square units
    Area of triangle D $= \frac{1}{2} \times 2 \times 3 =$
    3 square units
    Area of triangle E $= \frac{1}{2} \times 3 \times 3 =$
    4.5 square units

So B, C and D [2]
[1 mark for any correct area]

5. a) Area of triangle X $= \frac{1}{2} \times a \times h = \frac{1}{2}ah$ [1]
    b) Area of triangle Y $= \frac{1}{2} \times b \times h = \frac{1}{2}bh$ [1]
    c) Area of trapezium $= \frac{1}{2}ah + \frac{1}{2}bh$
       $= \frac{1}{2}(a + b)h$ [1]

6. a) 31.4×3 [1]
       $= 94.2\,\text{cm}$ [1]
    b) $(31.4 \div 2) + 10$ or $15.7 + 10$ [1]
       $= 25.7\,\text{cm}$ [1]

7. a) Area $= \pi r^2$, Area $= \pi \times 5^2$ [1]
       $= 78.5\,\text{cm}^2$ [1]
    b) Area of large square =
       $10 \times 10 = 100\,\text{cm}^2$ or
       Area of right-angled triangle =
       $\frac{1}{2} \times 5 \times 5 = 12.5\,\text{cm}^2$ [1]
       Area of small square $= 100 - 4 \times 12.5$
       or $4 \times 12.5 = 50\,\text{cm}^2$ [1]
       Unshaded area $= 78.5 - 50 = 28.5\,\text{cm}^2$ [1]

Pages 154–155
Statistics and Data

1. a) Mode $= 4$ [1]
    b) Putting numbers in order gives
       3, 4, 4, 4, ⑥, 8, 8, 13, 13
       So median is 5th value $= 6$ [1]
    c) Mean =
       $(6 + 3 + 8 + 4 + 13 + 8 + 4 + 4 + 13) \div 9$ [1]
       $= 7$ [1]
    d) Range = largest − smallest $= 13 - 3 = 10$ [1]

2.

Type of vehicle	Tally	Frequency
Car or van	JHT JHT JHT JHT II	22
Lorry or bus	JHT JHT I	11
Bicycle	II	2
Other	JHT III	8	[3]

[1 mark for correct tally column; 1 mark for three correct frequencies]

3. Total number of pencils is
   $8 + 12 + 11 + 15 + 9 + 11 = 66$
   There are 6 students, so each person should
   have $66 \div 6$ [1]
   $= 11$ pencils [1]

4. Total amount spent $= (4 \times £40) + £60$ or
   £160 + £60 or £220 [1]
   Mean amount $= £220 \div 5$ [1]
   $= £44$ [1]

5. a)

	Stall	Circle	Balcony	Total
Adult	240	**153**	192	**585**
Child	160	147	108	**415**
Total	**400**	300	**300**	1000

[2]

   [1 mark for at least two correct values]
   b) Adult tickets – child tickets $= 585 - 415$
   or $(240 - 160) + (153 - 147) +$
   $(192 - 108)$ or $80 + 6 + 84$ [1]
   $= 170$ [1]
   c) $585 \times 25 + 415 \times 20$ or $14\,625 + 8300$ [1]
   $= £22\,925$ [1]

6. a) Modal class has the greatest frequency,
   so 0–9 [1]
   b) Over 200 texts is an outlier, so mean is
   not suitable [1]

Pages 156–157

Decimals

1. a) $10 \times 10 \times 10 = 10^3$ [1]
   b) $10 + 10 + 10 = 30$ [1]
   c) $1 \times 10 \times 10 = 10^2$ [1]
   d) $1 = 10^0$ [1]

2.

Power of 10	10^{-1}	10^{-3}	10^{-2}	10^{-4}
Fraction	$\frac{1}{10}$	$\frac{1}{1000}$	$\frac{1}{100}$	$\frac{1}{10\,000}$
Decimal	0.1	0.001	0.01	0.0001

[4]

[1 mark for each correct column]

3. a) $18.34 - 18 = 0.34$ [1]
   b) $18.34 - 18.1 = 0.24$ [1]
   c) $18.34 - 18.04 = 0.3$ [1]
4. a) $402.836 - 129.325 = 273.511$ [3]
   [1 mark for 402.836; 1 mark for 273.511]
   b) $824.175 - 691.426 = 132.749$ [3]
   **[1 mark for 824.175 or 691.426; 1 mark for
   824.175 and 691.426]**
5. a) 100 [1]
   b) 0.1 [1]
   c) 1107 [1]
6. a) 2.7 [1]
   b) 2.7 [1]
   c) 2.7 [1]
   d) 0.083 [1]
   e) 0.083 [1]
   f) 0.083 [1]

7. $5 \times (13 \times 2.6 - 13 \times 0.6) =$
   $5 \times (13 \times (2.6 - 0.6))$ [1]
   $= 5 \times 13 \times 2 = 130$ [1]
   $10^2 + 10 \times 3 = 100 + 30 = 130$ [1]
8. a) 80 [1]
   b) 80 [1]
   c) 300 [1]
   d) 300 [1]
   e) 5500 [1]
9.

	1 s.f.	2 s.f.	3 s.f.	4 s.f.
1407	1000	1400	1410	1407
2999	3000	3000	3000	2999

[4]

   [1 mark for each correct column]
10. a) 40×4 or 39×4 [1]
    $= 160\,$cm or $= 156\,$cm [1]
    b) 40×40 [1]
    $= 1600\,$cm^2 [1]
11. a) $12.8 \times 1.5 = 2 \times 6.4 \times 1.5$ or 2×9.6 [1]
    $= 19.2$ [1]
    b) $6.4 \times 2.5 = 6.4 \times 1.5 + 6.4$ or $9.6 + 6.4$ [1]
    $= 16$ [1]

Pages 158–159

Algebra

1. a) $4a$ [1]
   b) b^3 [1]
   c) $2x + 6y$ [1]
   d) $x^2 - 3y$ [2]
   [1 mark for each term]
2. a) $9x + 5y - 6x + 7y = 3x + 12y$ [2]
   [1 mark for each]
   b) $7w + 5z - w - 7z = 6w - 2z$ [2]
   [1 mark for each]
3. $4 \times 3 + 6 \times 2 - 2 \times 1$ or $12 + 12 - 2$ [1]
   $= 22$ [1]
4. a) $360 \div 5 = 72°$ [1]
   b) $360 \div 10 = 36°$ or $360 \div 36 = 10$
   (decagon has 10 sides) [1]

5. There are 24 hours in a day or for example
   when $d = 1$, $h = 24$
   $h = 24d$ [1]
6. a)

×	3a	4
5a	$15a^2$	20a
4	12a	16

[3]

   [1 mark for one correct product]
   b) $15a^2 + 20a + 12a + 16 = 15a^2 + 32a + 16$ [1]
7. a) $3x + 24$ [1]
   b) $30x + 240$ [1]
   c) $3ax + 24a$ [1]
   d) $3abx + 24ab$ [1]
8. a) $a^2 + 3a + 4a + 12$ [1]
   $= a^2 + 7a + 12$ [1]
   b) $b^2 + 6b - 2b - 12$ [1]
   $= b^2 + 4b - 12$ [1]
   c) $c^2 + 5c + 5c + 25$ [1]
   $= c^2 + 10c + 25$ [1]
9. He has not collected like terms and written
   out the answer as an expression.
   $(3x + 4)(2x + 5y + 6)$
   $= 6x^2 + 8x + 15xy + 20y + 18x + 24$
   $= 6x^2 + 26x + 15xy + 20y + 24$ [1]

Pages 160–161

3D Shapes: Volume and Surface Area

1.

Name	Number of faces	Number of edges	Number of vertices
Cuboid	6	12	8
Tetrahedron	4	6	4
Cylinder	3	2	0
Triangular prism	5	9	6

[4]

   [1 mark for each correct row]
2. a) 9 [1]
   b) 16 [1]
   c) 9 [1]
3.

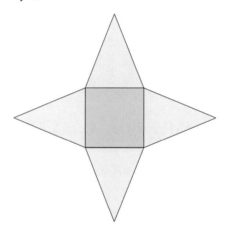

[1]

4. Surface area = $2 \times (10 \times 3) + 2 \times (3 \times 4) + 2$
   $\times (4 \times 10)$ or $2 \times 30 + 2 \times 12 + 2 \times 40$ or
   $60 + 24 + 80$ [1]
   $= 164 \text{ cm}^2$ [1]
   Volume = $10 \times 3 \times 4$ [1]
   $= 120 \text{ cm}^3$ [1]

5. a) 8 [1]
   b) 10 [1]

6. **a)**

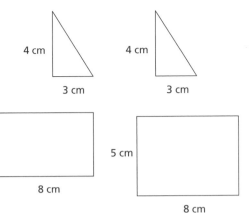

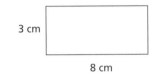

[4]

[1 mark for each unique face]

b) Area of triangle $\frac{1}{2} \times 3 \times 4$ or $6\,cm^2$ **[1]**

Area of a rectangle is 4×8 or 32 or 5×8 or 40 or 3×8 or 24 **[1]**

Total surface area = $(\frac{1}{2} \times 3 \times 4)$ + $(\frac{1}{2} \times 3 \times 4) + (4 \times 8) + (5 \times 8) + (3 \times 8)$ or

$6 + 6 + 32 + 40 + 24$ **[1]**

$= 108\,cm^2$ **[1]**

7. **a)** Volume = $\pi \times 3^2 \times 8$ or 72π or 226.19…

[1]

$= 226.2\,cm^3$ **[1]**

b) Total surface area = $\pi \times 3^2 \times 2 + 2 \times \pi \times 3 \times 8$ or $18\pi + 48\pi$ or 66π or 207.34… **[1]**

$= 207.3\,cm^2$ **[1]**

c) $25 \div 8 = 3.125$, so 3 **[1]**

Pages 162–163

Interpreting Data

1. **a)**

Number of students	30	1	3	4	6	7	10
Angle size	360°	12°	36°	48°	72°	84°	120°

[3]

[1 mark for 12°; 1 mark for at least three more correct angles]

b)

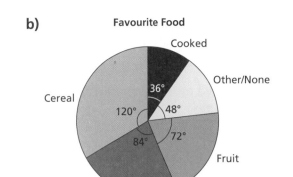

Favourite Food

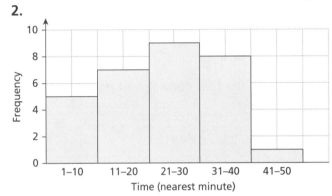

[2]

[1 mark for at least two correct sectors or for all sectors correct but without labels]

2.

[2]

[1 mark for at least two correct bars]

3. **a)** Negative **[1]**

b) No or zero **[1]**

c) Positive **[1]**

4. **a)** Year 8 **[1]**

b) There are more than 150 students in Year 11. **[1]**

5. 360° represents 24 students, so 360° ÷ 24 = 15° represents 1 student **[1]**

So 45° ÷ 15° = 3 students get 0 homework

60° ÷ 15° = 4 students get 1 homework

90° ÷ 15° = 6 students get 2 homeworks

165° ÷ 15° = 11 students get 3 homeworks

[1]

Total number of homeworks = $0 + (4 \times 1)$ + $(6 \times 2) + (11 \times 3) = 4 + 12 + 33 = 49$ **[1]**

Mean number of homeworks = $49 \div 24 = 2.04$ **[1]**

6. Range for farm A is $9 - 5 = 4$ years and the range for farm B is $11 - 4 = 7$ years **[1]**

This appears to support Ted but 4 is an outlier and does not seem generally representative of the animals on farm B, so without that single value the range for farm B would be $11 - 8 = 3$ years. So farm B looks to have more consistent ages. **[1]**

Pages 164–165

Fractions

1. Any fraction such that $\frac{1}{3}$ (0.33...) < fraction < $\frac{1}{2}$ (0.5)

   For example, $\frac{2}{5}$ or $\frac{5}{12}$ **[1]**

2. $\frac{6}{16}$, $\frac{12}{32}$ and $\frac{24}{64}$ **[2]**

   [1 mark for two correct and no incorrect fractions]

3. a)

   [1 mark for each grid]

   b) $\frac{5}{8} > \frac{3}{5}$ **[1]**

4. a) $\frac{2}{4} = \frac{1}{2}$ **[1]**

   b) $\frac{4}{5}$ **[1]**

   c) $\frac{4}{6} = \frac{2}{3}$ **[1]**

   d) $\frac{3}{8} + \frac{2}{8}$ **[1]**

   $= \frac{5}{8}$ **[1]**

   e) $\frac{7}{9} - \frac{6}{9}$ **[1]**

   $= \frac{1}{9}$ **[1]**

   f) $\frac{27}{30} + \frac{6}{30} - \frac{20}{30}$ **[1]**

   $= \frac{13}{30}$ **[1]**

5. $\frac{1}{2} \times \frac{3}{5} = \frac{3}{10}$
   [2]

   [1 mark for correct shading of rectangle]

6. a) $5 \times \frac{1}{3} = \frac{5}{3}$ **[1]**

   b) $\frac{1}{3} \times 4 = \frac{4}{3}$ and $1\frac{2}{3} = \frac{5}{3}$

   So $\frac{1}{3} \times 4 < 1\frac{2}{3}$ **[1]**

   c) $\frac{1}{2} \times \frac{1}{3} = \frac{1}{6}$

   So $\frac{1}{5} > \frac{1}{2} \times \frac{1}{3}$ **[1]**

   d) $\frac{1}{3} \times \frac{1}{2} = \frac{1}{6}$

   So $\frac{1}{3} \times \frac{1}{2} < \frac{2}{5}$ **[1]**

7. a) $\frac{1}{12}$ **[1]**

   b) $\frac{5}{12}$ **[1]**

   c) $\frac{1}{2}$ **[1]**

   d) $\frac{2}{5}$ **[1]**

8. a) $\frac{14}{9} \times 234 = 2 \times \frac{7}{9} \times 234$ or 2×182 **[1]**

   $= 364$ **[1]**

   b) $\frac{7}{3} \times 234 = 3 \times \frac{7}{9} \times 234$ or 3×182 **[1]**

   $= 546$ **[1]**

   c) $\frac{7}{18} \times 234 = \frac{1}{2} \times \frac{7}{9} \times 234$ or $\frac{1}{2} \times 182$ **[1]**

   $= 91$ **[1]**

9. $2\frac{1}{2} - \frac{3}{4}$ or 2 hours 30 minutes – 45 minutes

   $= 1$ hour 45 minutes **[1]**

   $= 1\frac{3}{4}$ hours **[1]**

Pages 166–167

Coordinates and Graphs

1.

Equation	Gradient	Coordinates of y-intercept
$y = -2x + 6$	-2	$(0, 6)$
$y = -2$	0	$(0, -2)$
$y = -8x$	-8	$(0, 0)$
$y = -9 + 0.4x$	0.4	$(0, -9)$
$-0.5x - 0.6 = y$	-0.5	$(0, -0.6)$

[5]

[1 mark for each correct row]

2. **a)** A has equation $y = x + 2$ [1]
   b) B has equation $x + 2 = 0$ [1]
   c) C has equation $y = -x + 2$ [1]
   d) D has equation $y = 2$ [1]
3. **a)** Any point on the line, e.g. $(-2, -2)$,
   $(0, -1)$, $(2, 0)$, $(4, 1)$, $(6, 2)$ [1]
   b) Any point below the line, e.g. $(2, -2)$ [1]
   c) Any point above the line, e.g. $(1, 2)$ [1]

4. **a)** $y = 2x + 3$ [2]
   [1 mark for finding a correct relationship for one point, e.g. $y = x + 7$ for the point $(4, 11)$]
   b) Yes, $y = 2x + 3$ gives a straight line [1]
5. D [1]
6. Correct equation is $x = y - 6$ [2]
   [1 mark for checking the coordinates of a point, e.g. when $x = -4$, $y = -10$]
7. **a)**

x	-3	-2	-1	0	1	2	3
y	7	2	-1	-2	-1	2	7

[2]
   [1 mark for any two correct values]
   b)

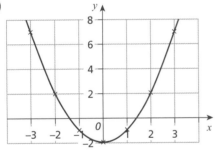

[2]
   [1 mark for at least four correctly plotted points]

Pages 168–169

Angles

1. $a = 180° - 150° = 30°$ (angles on a straight line are supplementary (add up to 180°)) [1]
   $b = 360° - 165° - 45° = 150°$ (angles at a point add up to 360°) [1]

$c = 67°$ (vertically opposite angles are equal) [1]
$d = 180° - 67° = 113°$ (angles on a straight line are supplementary (add up to 180°)) [1]
$e = 180° - 38° - 32° = 110°$ (angles in a triangle are supplementary (add up to 180°)) [1]
$f = 180° - 90° - 66° = 24°$ (angles in a triangle are supplementary (add up to 180°)) [1]
$g = 360° - 80° - 120° - 45° = 115°$ (angles in a quadrilateral add up to 360°) [1]

2. **a)** c, e and g [1]
   b) d, f and h [1]
   c) c and e or d and f [1]
   d) a and e or b and f or c and g or d and h [1]

3. $b = 180° - 35° = 145°$ (angles on a straight line add up to 180° / are supplementary) [1]
   $c = 145°$ (alternate to b) [1]
4. $a = 180° - 100° = 80°$ [1]
   $b = 45°$ [1]
   $c = 100°$ [1]
   $d = 180° - 45° = 135°$ [1]
   $e = 135°$ [1]
   $f = 180° - 80° - 45° = 55°$ [1]
5. **a)** Rhombus (four equal sides) [1]
   b) Kite (adjacent pairs of equal sides) [1]
   c) Arrowhead or delta (adjacent pairs of equal sides and a reflex angle) [1]
6. Octagon split into six triangles so sum of interior angles = $6 \times 180°$ [1]
   = 1080° [1]

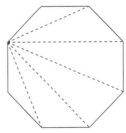

Pages 170–171

Probability

1.

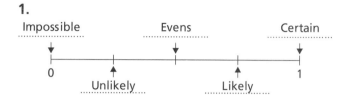

[4]

[3 marks for three correct; 2 marks for two correct; 1 mark for one correct]

2. a) Odd numbers are 1, 3, 5, 7, 9

    so $\frac{5}{10}$ or $\frac{1}{2}$ [1]

    b) Prime numbers are 2, 3, 5, 7

    so $\frac{4}{10}$ or $\frac{2}{5}$ [1]

    c) Multiples of 3 are 3, 6, 9 so seven

    numbers are **not** multiples of 3 so $\frac{7}{10}$ [1]

3. No. If the dice is fair, it is equally likely

    to land on 1, 2, 3, 4, 5 or 6 [1]

4. Sample space or listing of scores that shows

    four ways of scoring 5 [1]

Scores		Second card			
	+	1	2	3	4
First card	1		3	4	5
	2	3		5	6
	3	4	5		7
	4	5	6	7	

Most likely score is 5 [1]

As both cards are taken together, it is not possible to have same card twice, e.g. 1 and 1

5. a) $\frac{1}{6}$ [1]

    b) $\frac{2}{6}$ or $\frac{1}{3}$ [1]

    c) $\frac{3}{6}$ or $\frac{1}{2}$ [1]

    d) $\frac{1}{2} \times 50 = 25$ [1]

6. a) A = {3, 6, 9, 12, 15, 18, 21} [1]

    b) B′ = {12, 15, 21} [1]

    c) A ∪ B = {1, 2, 3, 6, 9, 12, 15, 18, 21} [1]

    d) Factors of 18 [1]

    e) $\frac{4}{9}$ [1]

7. a) (HH), HT, TH, TT [1]

    b) $\frac{1}{4}$ [1]

Pages 172–173

Fractions, Decimals and Percentages

1.

Fraction	Decimal	Percentage
$\frac{7}{10}$	0.7	70%
$\frac{2}{5}$	0.4	40%
$\frac{75}{100}$ or $\frac{3}{4}$	0.75	75%

[3]

[1 mark for each correct row]

2. a) $\frac{2}{3} = 0.666...$

    So $0.65 < \frac{2}{3}$ [1]

    b) $\frac{4}{5} = 0.8$ and $\frac{2}{3} = 0.666...$

    So $\frac{4}{5} > \frac{2}{3}$ [1]

    c) $\frac{4}{5} = 0.8$

    So $\frac{4}{5} > 0.65$ [1]

3. $-\frac{1}{3}$ -0.3 0.1 $\frac{1}{6}$ [2]

    [1 mark for any three in correct order (ignoring the fourth value)]

4. a) 0.4 [1]

    b) $\frac{4}{10}$ or $\frac{2}{5}$ [1]

5. $\frac{2}{3}$ of 60 $= \frac{2}{3} \times 60$ [1]

    $= 40$ [1]

6. 10% of 40 kg = 4 kg or 30% of 40 kg =

    $\frac{3}{10} \times 40$ [1]

    $= 12$ kg [1]

7. **a)** 10% of £40 = £4 or £40 + 3 × £4 or

1.3 × £40 [1]

= £52 [1]

b) 10% of £30 = £3 or £30 + 4 × £3 or

1.4 × £30 [1]

= £42 [1]

c) 10% of £50 = £5 or £50 − 2 × £5 or

0.8 × £50 [1]

= £40 [1]

d) 10% of £60 = £6 or £60 − 7 × £6 or

0.3 × £60 [1]

= £18 [1]

8. **a)** Boxes joined as follows:

Increase £40 by 15%   to   £40 × 1.15 [1]

Increase £15 by 40%   to   £15 × 1.4 [1]

Decrease £40 by 15%   to   £40 × 0.85 [1]

Decrease £15 by 40%   to   £15 × 0.6 [1]

b) £40 × 1.5   Increase £40 by 50% [1]

£40 × 0.6   Decrease £40 by 40% [1]

9. $\frac{15}{25} \times 100$ or 15 × 4 [1]

= 60% [1]

10. **a)** 1% of £500 = £5 or

4% of £500 = $\frac{4}{100} \times 500$ [1]

= £20 [1]

b) £500 + 3 × £20 [1]

= £560 [1]

Pages 174–175

Equations

1. **a)** $a = 11 - 7$, $a = 4$ [1]

b) $b = 15 + 9$, $b = 24$ [1]

c) $c = 24 \div 4$, $c = 6$ [1]

d) $d = 6 \times 6$, $d = 36$ [1]

e) $8e = 13 + 3$, $8e = 16$ [1]

$e = 2$ [1]

f) $5f = 21 - 1$, $5f = 20$ [1]

$f = 4$ [1]

2. **a)** Hearts are lighter as more circles with hearts, so circles heavier. [1]

b) Mass in first box = 4 × 2 + 6 × 3 =

8 + 18 = 26 g [1]

Mass of hearts = 26 − 4 × 5 or 6 g [1]

So each heart has mass 6 ÷ 3 = 2 g [1]

3. **a)** $5x - 2 = 28$ [1]

$5x = 30$, $x = 6$ [1]

b) Using Theo's number for example

$(6 - 2) \times 5 = 20$ [1]

$\neq 28$ Not correct [1]

4. **a)** $2x = 7 - 1$ or $2x = 6$ [1]

$x = 3$ [1]

b) $3x - 2x = 4 - 4$ [1]

$x = 0$ [1]

c) No solution [1]

d) $-4 = 7x - 5x$ or $-4 = 2x$ [1]

$x = -2$ [1]

5. **a)** All have same solution [1]

b) $9x + 1 = 4x + 21$ [1]

So $5x + 1 = 21$, $5x = 20$, $x = 4$ [1]

6. **a)** $3x + x = 21 - 5$ [1]

$4x = 16$ [1]

$x = 4$ [1]

b) $4y + 7 = 9 \times 3$, $4y + 7 = 27$ [1]

$4y = 27 - 7$, $4y = 20$ [1]

$y = 5$ [1]

c) $6z - 2 = 4z + 8$ [1]

$6z - 4z = 8 + 2$, $2z = 10$ [1]

$z = 5$ [1]

7. $3x - 8 + x + x = 22$ [1]

$5x - 8 = 22$, $5x = 22 + 8$, $5x = 30$ [1]

$x = 6$ [1]

Pages 176–177

Symmetry and Enlargement

1. Example answers: 90° clockwise rotation about C or 90° anticlockwise rotation about B or 180° rotation about the midpoint of BC (clockwise or anticlockwise) [2]

[1 mark for one correct description]

2. a)

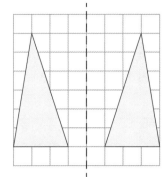

[1]

b)

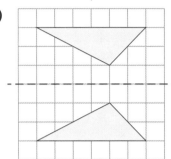

[1]

c)

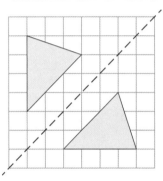

[2]

[1 mark for two vertices in correct position]

d)

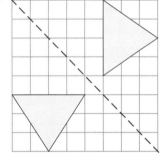

[2]

[1 mark for two vertices in correct position]

3. Order of rotational symmetry 4, 2, 2, 4, 4, 1

[2]

[1 mark for three, four or five correct]
The parallelogram does not have line symmetry. [1]

4. a) Scale factor = 14 ÷ 7 = 2 [1]

$a = 4 \times 2 = 8$ [1]

b) Scale factor = 4 ÷ 2 = 0.5 [1]

$b = 7 \times 0.5$ or $7 ÷ 2 = 3.5$ [1]

c) Ratio of perimeters = 4 : 8 : 2 = 2 : 4 : 1 [1]

5. a) False, the scale factor of enlargement is 9 ÷ 6 = 1.5 [1]

b) True [1]

c) False, $x = 40°$ as angles are preserved during enlargements [1]

6. 4.5 m = 450 cm, 450 ÷ 18 [1]

= 25 cm [1]

Pages 178–179
Ratio and Proportion

1. 6 : 10 = 3 : 5 [1]

2. a) Horizontal multiplier is × 2; vertical multiplier is × 5 [1]

Missing number is 20 × 2 or 8 × 5 = 40 [1]

b) Horizontal multiplier is × 1.5; vertical multiplier is × 2.5 [1]

Missing number is 15 × 1.5 or 9 × 2.5 = 22.5 [1]

c) Horizontal multiplier is × 1.25; vertical multiplier is × 0.75 [1]

Missing number is 6 × 1.25 or 10 × 0.75 = 7.5 [1]

3. a) Vertical multiplier: 1.2 [1]

£30 × 1.2 = £36 [1]

b) Vertical multiplier: 0.75 [1]

10 kg × 0.75 = 7.5 kg [1]

c) Vertical multiplier: 1.25 [1]

40 km × 1.25 = 50 km [1]

4. a) 8 parts is 24 squares, so 1 part = 3 squares [1]

3 red squares, 6 green squares and 15 green squares [1]

b) Not possible as 24 ÷ (4 + 3 + 3) is not an integer [1]

5. a) $\frac{2}{5}$ [1]

b) 150 ÷ 5 × 3 [1]

= 90 [1]

6. 3400 × 2 [1]

= 6800 km [1]

7. a) Ratio of steps is 2 : 3, $240 \times \frac{3}{2}$ [1]

= 360 [1]

b) $960 \times \frac{2}{3}$ or $640 \times \frac{3}{2}$ [1]

$960 \times \frac{2}{3} = 640$ or $640 \times \frac{3}{2} = 960$,

so correct [1]

8. 3 parts = 6 pens [1]

So, 1 part is 2 pens and 4 parts is 8 pens

So there are 8 blue pens [1]

9. $250 \div 4 \times 6$ [1]

= 375 g [1]

Pages 180–181

Real-Life Graphs and Rates

1. a)

Pounds (£)	0	10	20	30	40	50
Euros (€)	0	12	24	36	48	60

[2]

[1 mark for two correct conversions]

b)

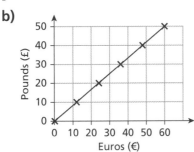

[2]

[1 mark for correct plotting of at least two points]

c) €42 (allow €41 to €43) [2]

[1 mark for attempt to read from £35]

d) For example, €900 = (£50 + £25) × 10 [1]

= £750 [1]

2. a) £17.50 (allow reading between £17 and £18) [1]

b)

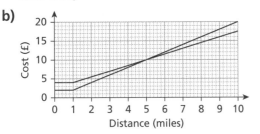

[2]

[1 mark for each correct line]

c) Terrier Taxis [1]

3. **a)** 30 minutes [1]

b) 125 × 2 = 250 km [2]

[1 mark for 125]

c) 4.30 pm [1]

d) (125 − 75) ÷ 2 or 50 ÷ 2 [1]

= 25 km/h [1]

4. 190 ÷ 60 [1]

= 3.16… so over 3 hours [1]

Alternative method: 60 × 3 [1]

= 180 miles, so 10 miles remaining after 3 hours [1]

5. Volume = $\frac{\text{mass}}{\text{density}} = \frac{60}{7.5}$ [1]

= 8 cm^3 [1]

Pages 182–183

Right-Angled Triangles

1. a) $6^2 + 8^2 = 36 + 64 = 100$ [1]

$\sqrt{100} = 10$ cm [1]

b) $9^2 + 5^2 = 81 + 25 = 106$ [1]

$\sqrt{106} = 10.3$ cm (1 d.p.) [1]

c) $13^2 − 12^2 = 169 − 144 = 25$ [1]

$\sqrt{25} = 5$ cm [1]

d) $10^2 − 7^2 = 100 − 49 = 51$ [1]

$\sqrt{51} = 7.1$ cm (1 d.p.) [1]

2. a) $8^2 + 15^2 = 64 + 225 = 289$ [1]

$= 17^2$, so right-angled [1]

b) $7^2 + 20^2 = 49 + 400 = 449$ [1]

$\neq 21^2$, so not right-angled [1]

c) $2.5^2 + 6^2 = 6.25 + 36 = 42.25$ [1]

$= 6.5^2$, so right-angled [1]

3. Using Pythagoras' Theorem $l^2 = 4^2 + 3^2$, where l is the slant height [1]

$l^2 = 16 + 9$, $l^2 = 25$ [1]

$l = 5$ units [1]

Perimeter = 8 + 5 + 5 = 18 units [1]

4. a) 0.95 [1]

b) 0.95 [1]

c) 3.28 [1]

d) 10.07 [1]

e) 60° [1]

f) 25° [1]

5. a) $a = 10 \sin 40°$ [1]

= 6.428 cm [1]

b) $b = 5 \sin 40°$ [1]

= 3.214 cm [1]

c) $\sin 40° = \dfrac{12.856}{c}$ [1]

$c = 20$ cm [1]

6. a) $\sin x = \dfrac{3}{8}$ [1]

$x = \sin^{-1}(\dfrac{3}{8}) = 22.0°$ [1]

b) $\cos x = \dfrac{6}{12}$ [1]

$x = \cos^{-1}(\dfrac{6}{12}) = 60.0°$ [1]

c) $\tan x = \dfrac{9}{5}$ [1]

$x = \tan^{-1}(\dfrac{9}{5}) = 60.9°$ [1]

Mixed Test-Style Questions
Pages 184–192
No Calculator Allowed

1. a) $1000 = 2^3 \times 5^3$ [2]

[1 mark for choosing 2, 2, 2, 5, 5, 5]

b) HCF $= 2^3 \times 5$ [1]

= 40 [1]

2. a) $15 \times 243 = 14 \times 243 + 243$ or $3402 + 243$ [1]

= 3645 [1]

b) $87 \times 24 + 87 \times 6 = 87 \times 30$ [1]

= 2610 [1]

3. a) False [1]

b) True [1]

c) False [1]

4. a) Incorrect [1]

b) Correct [1]

c) Correct [1]

d) Correct [1]

5. D Number of tickets × 20 + 4 [1]

6. a) Correct [1]

b) Incorrect [1]

c) Correct [1]

7. a) $x + y = 5$ [1]

b) Any coordinates whose sum is 5, e.g. (2, 3) [1]

8. $\dfrac{7}{8} \times \dfrac{8}{9} = \dfrac{7}{9}$ (0.777…) and $\dfrac{8}{9} \times \dfrac{9}{10} = \dfrac{8}{10}$ (0.8)

So the second calculation gives the greater answer [2]

[1 mark for one correct calculation or conversion]

9. a) Any numbers in the ratio 5 : 4, e.g. 5 and 4 or 20 and 16 [1]

b) $18 \div 4 = 4.5$ [1]

10. 1026 [2]

[1 mark for two of 600 or 160 or 210 or 56 or one of 266 or 760 or 810 or 216]

11. a) $10a + b$ [2]

[1 mark for each term]

b) $4c + 32d + 5c - 35d = 9c - 3d$ [2]

[1 mark for correct expansion or $9c$ or $-3d$]

12. a) Reflection in the line $x = 2$ [1]

b) Reflection in the line $x = 0$ or y-axis [1]

c) Reflection in the line $y = 1$ [1]

13. a) (Pattern 4) $2 + 4 + 4 + 4 + 4 = 18$ [1]

b) $2 + 25 \times 4$ [1]

= 102 [1]

c) $(198 - 2) \div 4$ [1]

= 49, so pattern 49 [1]

14. a) $7 \times 3 - 5 = 16$ [1]

b) $(-5 + 5) \div 3 = 0$ [1]

15. $d = a$ (alternate angles) [1]

$e = c$ (alternate angles) [1]

$d + b + e = 180°$ (angles on a straight line are supplementary (or add up to 180°)) [1]

so $a + b + c = 180°$ [1]

Therefore, the angles of a triangle add up to 180° [1]

16. Example method. Use 8 cm as the base line; arc from one end of radius 6 cm; arc from other end of radius 3 cm; triangle completed through point of intersection of the arcs.

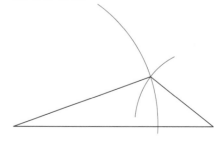

[2]

[1 mark for a correct baseline and one correct arc]

Alternative method using full circles:

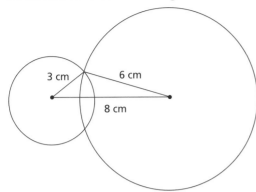

17. a) Any pair from 1 and 24 or 2 and 12 or 3 and 8 or 4 and 6 [1]
    b) A different pair from 1 and 24 or 2 and 12 or 3 and 8 or 4 and 6 [1]

18. a) $83 \times 25 = 8.3 \times 10 \times 2.5 \times 10$
    $= 20.75 \times 10 \times 10$ [1]
    $= 2075$ [1]
    b) $8.3 \times 12.5 = 8.3 \times 2.5 + 8.3 \times 10 = 20.75 + 83$ or $8.3 \times 2.5 \times 5 = 20.75 \times 5$ [1]
    $= 103.75$ [1]

19 a) Example answers: (1, 0), (0, 1), (–1, –1), (–1, 0), (–1, 1), etc. [1]
   b) Example answers: (–1, 3), (0, 2), (1, 1) [1]
   c) (2, 0) [1]

20. a) 20×0.5 or 18×0.5 [1]
    $= 10$ or $= 9$ [1]
    b) $(20 - 7) \div 2$ or $(19 - 7) \div 2$ [1]
    $= 6.5$ or $= 6$ [1]
    c) $(10 + 4) \div 0.5$ [1]
    $= 28$ [1]
    d) 30^2 [1]
    $= 900$ [1]

21. a) True [1]
    b) True [1]
    c) True [1]
    d) True [1]

22. $\frac{23}{5}$ [1]

23. (7, 5) [2]
    [1 mark for one correct coordinate or for drawing the correct translation]

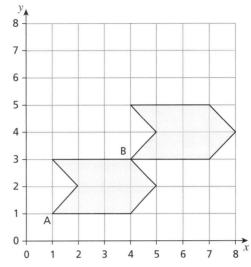

24. a) $\frac{4}{20}$ [1]
    $= \frac{1}{5}$ [1]
    b) $\frac{14}{20}$ [1]
    $= \frac{7}{10}$ [1]
    c) Not correct as, for example, there are 16 left so now P(square) $= \frac{3}{16} \neq \frac{1}{5}$ [1]

25. a) Mode $= 8$ [1]
    b) Putting numbers in order:
    2, 3, 4, 6, 7, 8, 8, 8, 9, 10 [1]
    Median $= 7.5$ [1]
    c) Mean $= (2 + 3 + 4 + 6 + 7 + 8 + 8 + 8 + 9 + 10) \div 10$ or $65 \div 10$ [1]
    $= 6.5$ [1]
    d) Range $= 10 - 2 = 8$ [1]

26. A greater proportion of adults attended the opera than the pop concert (60% compared with 30%) [1]

Pages 193–200

Calculator Allowed

1.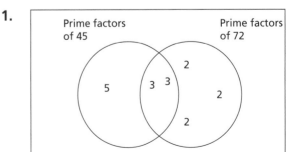
   [3]

[1 mark for each correct region]

2.   B, C and E   [2]

[1 mark for two correct and none incorrect]

3. $85° + 45°$ or $180 - 85° - 45°$ or $50°$ [1]

    $x = 130°$ [1]

4.   **a)** Any suitable answer, e.g. Amina had the shortest time, 11 seconds.   [1]

    b) Any suitable answer, e.g. Bob had the lowest total time 67 seconds (total for Amina was 68 seconds).   [1]

5.   Ashima and Rosie are both correct, but Ashima has not simplified the answer.   [2]

[1 mark if only one chosen]

6.   Not correct as sequence decreases by 3 each time or $3n$ would mean sequence is increasing   [1]

    So nth term $= -3n + 23$ or $3n + 17$ gives 20, 23, 26, 29, …   [1]

7.   **a)**

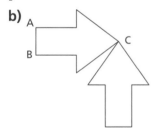

       [2]

[1 mark for correct orientation]

    b)

       [2]

[1 mark for correct orientation]

    c) Both images have the same orientation (point the same way)   [1]

8.   **a)** $1.74 + 1.52 + 0.13$ or 3.39 [1]

      $3.39 \div 2 = 1.695$ m   [1]

    b) $1.82 + 1.79 - 0.13 = 3.48$ [1]

      $3.48 \div 2 = 1.74$ m   [1]

9.   Each y-coordinate is 3 more than its x-coordinate, so equation is $y = x + 3$ [1]

10. $100\% - 60\% = 40\%$, so multiplier is 0.4 or $0.4 \times £280$ [1]

    $= £112$ [1]

11.   **a)** $8x = 65 + 7$ or $8x = 72$ [1]

      $x = 9$ [1]

    b) $x = 12 \times 2$, $x = 24$ [1]

12. $\sin 35° = \frac{x}{8}$ or $x = 8 \sin 35°$ [1]

    $= 4.588… = 4.6$ cm (to 1 d.p.)   [1]

13.   Volume is $200 \times 15 \times 4 = 12\,000$ cm^3 [1]

    Mass $= 7200$ g   [1]

    Density $= 7200 \div 12\,000$ [1]

    $= 0.6$ g/cm^3 [1]

14.   **a)** HK\$ 28   [1]

    b) $28 \times 3 \times 100$ or 28×300 [1]

      $=$ HK\$ 8400   [1]

    c) $£1 = 28 \div 3$ [1]

      $=$ HK\$ 9.33   [1]

15.   (diagonal^2 =) $8.5^2 + 3.2^2$ or 82.49 [1]

    diagonal $= \sqrt{8.5^2 + 3.2^2}$ or $\sqrt{82.49}$ [1]

    $= 9.08$ m or 9.1 m   [1]

16.   **a)** $2x(2x + 3)$ [2]

[1 mark for partial factorisation $2(2x^2 + 3x)$ or $x(4x + 6)$]

    b) $(x - 11)(x + 6)$ [2]

[1 mark for $(x + a)(x + b)$ where $ab = -66$ or $a + b = -5$]

17.   **a)** (HHH), HHT, HTH, THH, HTT, THT, TTH, TTT   [2]

[1 mark for at least five correct]

    b) $\frac{1}{8}$ [1]

18.   **a)**

x	-2	-1	0	1	2
y	-7	-5	-3	-1	1

       [2]

[1 mark for at least two correct values]

    b)

       [2]

[1 mark for at least two correct points plotted]

c) (0, −3) [1]

d) 2 [1]

19. **a)** Circumference of circle = $\pi \times 7.2$ or
Circumference of semicircle arc =
$\pi \times 7.2 \div 2$ [1]
= 11.3… cm [1]
Perimeter = 11.3 + 4 + 7.2 + 4 = 26.5 cm
(to 1 d.p.) [1]

b) Area of rectangle is $7.2 \times 4 = 28.8\,\text{cm}^2$ [1]
Area of circle is $\pi \times 3.6^2$ or Area of
semicircle is $\pi \times 3.6^2 \div 2 = 20.357…$ [1]
Area of shape is 28.8 + 20.357… =
$49.2\,\text{cm}^2$ (to 1 d.p.) [1]

20. Area of cross-section is $\frac{1}{2} \times 8 \times 6 = 24\,\text{cm}^2$ [1]
Volume of prism = $\frac{1}{2} \times 8 \times 6 \times 18$ or 24×18
= $432\,\text{cm}^3$ [1]

21. $36 \div 20 = 1.8$ m or $3600 \div 20$ [1]
= 180 cm [1]

22. **a)** $5 \times 4 \times £25.99$ [1]
= £519.80 [1]

b) Buying from the 3 m wide roll
$3.7 \times 3 = 11.1\,\text{m}^2$ or buying from the
4 m wide roll $2.9 \times 4 = 11.6\,\text{m}^2$ [1]
Buying from the 3 m wide roll needs
less carpet. [1]

23. $\frac{340}{2000} \times 100\%$ [1]
= 17% [1]

24. 371 [1]
= 370 (to 2 s.f.) [1]

25.

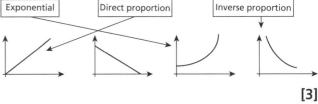

[3]

[1 mark for each correct match]

26. **a)**

Number of students	40	1	3	5	7	8	17
Angle size	360°	9°	27°	45°	63°	72°	153°

[3]

[2 marks for four correct angles; 1 mark for two correct angles]

b)

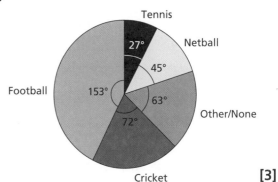

Favourite Sports

[3]

[2 marks for at least three correct sectors; 1 mark for at least one correct sector]

Acknowledgements

The authors and publisher are grateful to the copyright holders for permission to use quoted materials and images.

All images are © HarperCollins*Publishers* Limited

Every effort has been made to trace copyright holders and obtain their permission for the use of copyright material. The authors and publisher will gladly receive information enabling them to rectify any error or omission in subsequent editions. All facts are correct at time of going to press.

Published by Collins
An imprint of HarperCollins*Publishers*
1 London Bridge Street
London SE1 9GF

HarperCollins*Publishers*
Macken House, 39/40 Mayor Street Upper,
Dublin 1, D01 C9W8, Ireland

© HarperCollins*Publishers* Limited 2022

ISBN 9780008551445

First published 2022

10 9 8 7 6 5 4 3

British Library Cataloguing in Publication Data.

A CIP record of this book is available from the British Library.

Publisher: Clare Souza
Authors: Samya Abdullah, Rebecca Evans, Trevor Senior and Gillian Spragg
Videos: Anne Stothers
Project Management: Richard Toms
Cover Design: Sarah Duxbury and Kevin Robbins
Inside Concept Design: Sarah Duxbury and Paul Oates
Text Design and Layout: Jouve India Private Limited
Production: Emma Wood
Printed in Great Britain by Martins the Printers

This book contains FSC™ certified paper and other controlled sources to ensure responsible forest management.

For more information visit: www.harpercollins.co.uk/green